Getting Sued and Other Tales
of the Engineering Life

Getting Sued and Other Tales
of the Engineering Life

Richard L. Meehan

The MIT Press
Cambridge, Massachusetts
London, England

Third printing, 1984

This book was set in Linotron 202 Melior by Graphic Composition, Inc., and printed and bound by The Murray Printing Co. in the United States of America.

Library of Congress Cataloging in Publication Data

Meehan, Richard L.
  Getting sued, and other tales of the engineering
life.

  1. Meehan, Richard L.   2. Engineers—United
States—Biography.   I. Title.
TA140.M36A34   620′.0092′4   [B]   81–8152
ISBN 0–262–13167–6   AACR2

To My Mother and Father

Contents

Introduction

I was first introduced to calculus by a high school math teacher named Philbrick Bridgess, a dry and somber New Englander who would have fit well into a poem by Robert Frost. Mr. Bridgess had that rare ability for showing that the subject of his expertise was really ridiculously simple, and in this and other of his taciturn ways provided me with encouragement, which I badly needed in those forlorn adolescent days. I remember once after introducing us to the subject of integration, he passed out an aromatic mimeographed sheet filled with algebraic functions. "Integrate them, then differentiate 'em back again. That's a good habit," he said, pausing to look out the window over the school football field and track; a few years before, as seventh graders, we had made grubby computations of their perimeters and areas under his guidance. Then he made what was for him an uncharacteristically personal remark. "I had a roommate in college. He'd make up functions. Differentiate them. Integrate them back again. That way he'd know if he'd got them right. Seemed like a lot of work to me. That's why he's a big shot at GE and I'm a high school math teacher."

Mr. Bridgess's story, brief as it was, taught me something about math. And it taught me something about learning math, about teaching, and about Mr. Bridgess too. The calculus part of the lesson has faded from memory; no doubt today I would be unable to work the problems on that homework assignment. But I have not forgotten the other lessons in that laconic tale of the roommate.

This book aims in similar fashion to tell something about what it is like to become an engineer. I have approached the subject by telling some stories from my own experience, hoping to convey something of the flavor of the technological life, recognizing that supplementary vitamins and minerals, not to speak of fiber, can be found in other volumes such as the *Chemical Rubber Handbook*.

Probably there is a more systematic way to approach this subject, based, perhaps, on taxonomies of personality types, hierarchies of motives, schedules of life passages, and the like. It is true that we scientists and engineers generally prefer the systematic to the anecdotal, the Aristotelian to the Homeric. But there are occasions when the point may be made more readily with a story than a dissertation. For example, let us say that twenty years of experience has led me to believe there is nothing inherently more dehumanizing in technological enterprise than in medicine or law, that there is a lot more to a career in technology than Ohm's law and FORTRAN and mixing concrete, that people and morals and delight and misery are all waiting to confront you whether you're a computer engineer or a public-interest or corporate lawyer or a village headman in Thailand. I believe that. My experience over the last few years teaching graduate students has convinced me that this kind of message—and I think it is an important one—can be more readily transmitted to students anecdotally than systematically.

Certainly the details of everyone's experience vary, and I do not pretend that designers of computer microprocessors or professors of chemical engineering have the same stories I present here. Nonetheless I expect that most technologists, whatever their specialty, at one time leave home, just as I did, to join a professional tribe, there encountering people who have been similarly trained but have cultural backgrounds that differ greatly from their own. At one time or other, they will, like my man who bought Route 128, be introduced to financial pressures and temptations. They will experience, as I did, the confusion and panic that comes when one is caged with that different species, lawyers. They will discover that there are differences between homework problems and real problems; between clever analyses and clever designs; that organizations have a morality and will of their own; that bosses' or clients' goals are never entirely explicit or even rational; that there is always an unconscious gap between what people want and what they say they want.

Of course most of these lessons must be learned as well by accountants, military officers, architects, scientists, managers—anyone attempting to succeed in a profession. As Jacob Bronowski put it, "Whether our work is art or science or the daily work of society, it is only the form in which we explore our experience which is different." So in a broad sense, this book addresses the experience that befalls all of us between the time we decide on our work, at say age eighteen, and the time we finally understand something about what is really going on, at about age thirty.

Although the stories that follow are set in chronological order, the gaps in their sequence and the absence of characters from my personal life (certain of whom nonetheless cast shadows, which no adjustment of stage lighting will remove) disqualifies this collection of tales as even a journeyman's autobiography. However, to provide readers with some general orientation, I include a few brief paragraphs describing my employment history.

We lived a comfortable suburban life when I was a child, and certainly I had no need to take a job. However, when I was in the fifth grade, I was influenced by other boys who were able to present glamorously swaggering accounts of the joys of caddying. It is a marvel of the human imagination that one could find romance in spending the better part of a day carrying heavy leather bags of golf clubs for rich men, all for one dollar. However, one of the more comforting lessons of life is that in the end there are certain fixed quanta of pleasure and pain in almost any human activity or condition. The click of a perfect wood shot on a misty July morning; the unzipping of a strange leather pouch, pregnant with gleaming white balls and chrome yellow tees; the casual stroll up to the pin, carefully planned and perfectly executed so that one's shadow never falls between the ball and the hole; who can say that a slave's pleasures are not equal to a king's, a consultant's to a client's?

For several high school summers, I was a greengrocer's boy in the A&P. I whistled "Three Coins in the Fountain," split watermelons, and guarded the uppermost purple stratum of boxed cherries against assault by dangerous shoppers. After several summers as a produce clerk, I was promoted to checker, but I proved to be as inept a checker as I was an airplane pilot many years later, so I was demoted to the back room of the grocery department, where I slit cartons, stamped purple prices on cans, and speculated on the potentially bizarre disposition of horsemeat and sheets of Kleenex. I worked with men in their early twenties who hated themselves for loving the Great Atlantic and Pacific Tea Company, who swore they would get out even as they grew sick with desire for assistant store managerships. I was glad I did not need to worry about all that, for I possessed, and that summer read and reread, a single sheet of bond from MIT's admissions office, on which were typed the words that were going to carry me away from the A&P and into the world described in this book.

The summer after my freshman year at college, I took a job in the quality-control laboratory in a fish warehouse in South Boston. There I counted worms in cod filets and weighed fish sticks. My boss was a biochemist who derived an equation to predict the freshness of fish filets as a function of temperature and time. This was the first time I witnessed the transformation of a natural process into an equation, and I was much impressed by the magic and pleasure of this fundamental engineering operation.

The following summer, between sophomore and junior year, I worked as a construction inspector for a contractor at a BOMARC missile site. The BOMARC missiles were supposed to pop out of the ground on Cape Cod and fly up to Canada to intercept Russian bombers arriving from the North Pole. The BOMARC struck me as a sort of aerial Nash Rambler. I was always glad that no one found cause to fire them off, for I feared that they would as likely fly off in the wrong direction or simply blow up their little cinderblock houses. Several people told me that my work was quite good, and halfway through the summer I thought on this basis that I should ask for a raise. This resulted in my being instantly fired. Walking through a contractor's office after being canned, with the project manager shouting at me for being ungrateful for the opportunity of a summer job and the bookkeepers and secretary and clerk of the works turning pale and wishing they were somewhere else, is a ten-second experience in free enterprise that should not be missed by any one.

A full account of life as an engineering undergraduate deserves to be written. Here I've tried to capture only a small part of that experience, in chapter 1 and in part of "Haiti One More Time." Subsequent experience as a student-citizen-soldier, which students of the 1980s probably will once again contemplate with the same feelings that we experienced in those Berlin Wall days of the early 1960s, is described in "Confessions of a Military Engineer." I've not yet attempted to write an account of post-army life as a New York commuter, working for a large New York architect-engineering firm; that is a future project.

"A Dam for Lam Pra Plerng" describes two years of the engineering life in Thailand in the early 1960s. After Thailand, between the steamy afternoon in June that I left my dam at Lam Pra Plerng, bound for Calcutta, Peshawar, and Kabul, and that peculiar dreamlike morning, exactly two years later, when I arrived in Santiago, Chile (where my story picks up again), I spent a year in graduate school in London followed by a year back in Boston doing foundation engineering for structural engineer Bill LeMessurier. That year in Boston, 1967, I drilled soil borings in Cambridge argillite, Indian fish wiers, Beacon Hill, and, once, a buried cable carrying several thousand telephone conversations. Using the information gained from these boring results, I designed foundations for the high-rise buildings that were then changing the Boston skyline. Recently while visiting Copley Square with my daughter, I stopped to look at the construction joint between my favorite old building, the Boston Public Library, and Philip Johnson's new addition to it, a structure in which I have a special interest because I designed the daring foundation that supports it. The architect apparently chose not to follow the recommendation present in my report, which suggested that because both buildings are supported on two hundred feet depth of quaking, toothpaste-like Boston blue clay, they should be connected, if at all, with a joint that would permit two or three inches of differential play between them. Secretly I had faith in the efficacy of my Archimedean scheme of floating Johnson's massive new addition: carving out exactly enough soil from under it and designing the basements like the hull of a ship so that the weight of the excavated soil would equal the 300,000 ton weight of the structure and its heavy load of books. This way I figured the new addition would float, like the *Queen Elizabeth*, with tranquil equilibrium beside its prim neighbor. From the look of that cemented joint between old and new, unbroken today, I think the scheme worked. Far better, in any event, than a certain bungled effort at foundation engineering that nearly destroyed Trinity Church on the other side of the square.

My work in Boston was fun, but I soon departed Boston again for the Chilean Andes, an adventure described in "Snowbound on the Rio Pangal." But by the time I left my cabin in the bleak valley of the Pangal, the romance of life overseas had worn thin. More by accident than design, I settled on the San Francisco peninsula, where I have remained since. With the exception of an occasional pilgrimage to the tropics—one of which yielded the retaining wall described in "Haiti One More Time"—I have been content to tend my home gardens. In 1969, I joined with some geologists to start a consulting firm. Fortunately, "Getting Sued," the story that ends this book, did not end our consulting business, so the stories of other consulting and professorial events of the past few years await (to borrow an appropriately technological metaphor from Vladimir Nabokov) "the evaporation of certain volatiles and the melting of certain metals in my coils and crucibles."

The first of these stories, "A Dam for Lam Pra Plerng," was conceived at a moment of distress in the autumn of 1976. I kept a journal in those days, in which I find entered for September 26 the note, "Some ideas for essays. Thailand. How the dam was built. Luay, Vicha, Longmah, Nicom." This story and the ones that follow were written in various notebooks and on pieces of paper in a variety of settings, including Stapleton International Airport, the tomb of Leland Stanford, Jr., and the Oakland County courthouse. Final editing and polishing was done using the WILBUR text-editing system on an IBM 3033 computer.

These stories present people and events as I remember them. If I have sometimes seemed smug or intolerant in my portrayal of the character of those around me, let me say that none of them exhibit weaknesses or failings of which I myself have not been guilty at some time or in some context or degree.

This book is the result of the direct or indirect efforts of fifteen people, beginning with my grandmother, who loved words, and ending with my wife, Ruth, who maintained the effort with support and affection. The remaining dozen friends, business associates, and mentors provided the crucial alloys of tolerance, criticism, encouragement, and skill necessary to amplify into final copy what began as only the faintest of glows in my neural synapses. I thank them all.

*Getting Sued and Other Tales
of the Engineering Life*

Coming of Age at SAE

By the time I was ready to pick a college and a career in the late 1950s, I had learned that pursuit of romantic whims could bring woeful inconveniences, so I tried to be very cold-blooded and logical in my attack on the problem. After some deliberation, I concluded that my occupation had to satisfy three basic criteria.

First, I thought that it should be something for which I had some minimal inborn talent. Satisfaction of this rule required some understanding of the talents necessary for the work itself and also knowledge of my own abilities and weaknesses.

The know-your-job criterion was easy. My father was an engineer, and I knew that he knew about such matters as erector sets and concrete, refrigerators and film speeds. Indeed if (as it is said) child's play is a primitive form of work, I had undertaken my first engineering jobs on rainy winter days at age three, when, sitting on the floor of our Connecticut living room, listening to the season's hits, "Praise the Lord and Pass the Ammunition" and "White Christmas," I had built Incan tombs using my father's books. A text on steam engineering, which he still occasionally consults, served as a lintel that could be supported by twin pillars: one the U.S. Army Corps of Engineers *Field Manual*, 1917 edition (page 455: "A good mule has a soft kindly look in his eye which is difficult to describe but easily recognized"); the other a vermilion-colored schoolboy's edition of Plato's *Phaedra*, in Greek, heavily annotated in my father's neat hand.

The "know thyself" part of the rule was not so easy. When I was very young the women in my life had praised me for my capability in such enterprises as disassembling door knobs, which led me to think that I had special mechanical interests and talents suitable for engineering. "Just like his father," they said. Alas, a little feminine flattery can create a persistent delusion; I did not admit to myself that I had neither talents nor patience for mechanical devices until I was about twenty-five and had already become an engineer. Fortunately by that time I realized that it is not necessarily a disadvantage to lack a talent popularly considered fundamental to a profession; indeed I have since observed that the best of any trade are often those who discover that they are not very skilled at the day-to-day practice of its basic techniques. Perhaps from a sense of inadequacy and desperation they become leaders and innovators. This is not a cynical observation. Two of the most admirably successful engineers I have known failed their professional registration exams; one is a brilliant designer, and the other created a worldwide market for the engineering talents of his firm.

A second career-selection principle that I attempted to follow then, and which has served me well in many subsequent decisions, is what I call the short-line rule. I first applied this rule on registration day of my sophomore year at MIT. At that time I had not decided on a major so I spent the morning walking the corridors of the institute, visiting various of its twenty-two departments. Great crowds of students milled around some of the department headquarters. Several classmates I remembered from my freshman physics sections were lined up before the electrical engineering and physics departments. I recalled that most of them had seemed to know as much about physics as our instructors did. Clearly I was not in their league. But I had done well enough in my freshman math sections; to me calculus was a collection of magic tricks that made me want to laugh with delight. Perhaps I would become a math teacher. But when I arrived at the headquarters of the

3   Department of Mathematics, I found another long and slowly moving line. "A career in math is probably the same as that line," I thought to myself, and I knew at once that I did not want to be a mathematician.

The year was 1957. Civil engineering was much out of fashion in those days, and when I arrived at Building One, with its museum of model ships and laboratories of dusty heroic machines, there was no line at all in front of the civil engineering headquarters. In fact, the department head himself greeted me warmly and asked me to sit down for a little talk. "This department is in a bad way," I thought to myself afterward. But then later I asked myself why that might be. The answer, it seemed, was not that civil engineering was obsolescent but that it was unfashionable. The basic idea of the short-line rule is that you should buck the trend. So I wrote "civil engineering" in the appropriate space of my pink IBM card, dropped it at the registrar's office, and then went across the street to the Paradise Café for a beer. I have never regretted the decision.

For reasons that I do not understand, many otherwise logical and talented people, who in their hearts would like to attain some recognition for their professional accomplishments, defy the short-line rule by choosing careers in the same irrational way one might choose a pair of new shoes—on the basis of popular style. For example, when I was an undergraduate in the early 1960s *Life* magazine and the employment advertisements in the Sunday *New York Times* were full of enthusiasm over space exploration. Many students were excited by this publicity and worked toward careers in the aerospace industry. Predictably enough, the market became glutted with former space enthusiasts. Some of them ended up driving taxicabs and tending bar ten years later. A similar boom occurred a few years later in oceanography. There was a flurry of popular interest in the sea as a new frontier, which could be plundered in the old-fashioned mercantile manner, or, for the

4 counter-cultured, contemplated romantically in the new ecological mode. These reasons were enough to attract many students, and universities, which package and market their programs to students, not employers, met the demand. There was much disappointment when it became apparent that the job market for oceanographers was not much better than for moral philosophers. Were I a student today, I would be similarly careful to avoid that great hubbub of lawyers or to wander into that jobless wilderness of environmental studies.

My third principle for career selection derived from the hunch that ambitions in large part are fronts for fantasies and that it is surely much easier to design a career that will take you at least part of the way toward a fantasy if you recognize what it is. For example, a man might greatly admire an uncle who happened to be a teacher of biology, without recognizing that what he really admired about his uncle was his position at the head of the class, not his work as a biologist. Without really understanding this, our man might spend a misguided career as a research biologist when he might have been much happier as an English teacher or even a clergyman or an army officer. I have seen some people experience great difficulties in the consulting business because they have a mistaken idea, usually founded on their admiration of some prominent academic mentor, that a consultant is a sort of oracle whom clients seek out for advice. To be such a person—that is the dream. The reality of the situation, that consulting is a highly competitive business requiring sales skills and attitudes, that clients generally prefer tried-and-true solutions, not revolutionary paradigms, does not square with their fantasy. Hence, though such people may do an excellent job of emulating their mentors, they may have great difficulties as consultants because they lack the interest, the persistence, or the attitudes necessary to sell and provide professional services on the open market.

5   So it happened that because I knew the business and thought I might be good at it, because it was not in fashion, and because it was a practical trade, a means to an end, I went off to MIT to be an engineer.

Rush week began at the twenty-five fraternities of MIT in mid-September, two weeks before the start of fall classes. It was a week of breakfasts, brunches, lunches, teas, cocktails, dinners, and parties. Behind this festivity lurked a serious purpose: the recruitment of new members, the pledges, from MIT's incoming freshman class. During rush week every fraternity brother was expected to be alert for new material, to "bird-dog" the ideal "neat guy" who fit the fraternity's image of itself. The younger brothers, the sophomores, were on the front line; they were to seek out promising candidates and introduce them to the older members. Then late in the evening, before the assembled brotherhood, they nominated their discoveries for membership, usually with the endorsement of one of the older brothers. If their candidate was not blackballed, the same two-man team then made a bid to the new man the next day, attempting by pressure, flattery, all the usual sales techniques, to persuade the candidate to accept the pledge pin. The pin, if accepted, was presented to the new pledge in a joyous celebration that lasted forty-five seconds, after which the new pledge was pressed into service as a recruiter.

Rush week was serious and competitive business, for the economic survival and social success of the fraternity depended on the annual recruitment of a good pledge class. After rush week there might be a few more opportunities, scattered nuggets missed in the first sluice. But a fraternity that slipped up in rush week was in trouble, and if your house had a great pledge class, you learned from the older brothers how to shake your head sympathetically and say it was too bad about the Delts, they were really up the creek this year, in just the same style that, twenty years later, you'd talk about a business competitor who lost a big contract. In this and other ways, fraternities prepared young men for the American way of business. "Fraternities are probably the greatest bastion we have where we can develop leaders to take care of the protection of the Republic and our way of life," Barry Goldwater once said. Norman Vincent Peale put it even more simply: "The fraternity makes men."

I do not recall when the desire for fraternity life first came upon me, but by the late summer of 1957, with my enrollment as an MIT freshman only a month off, the idea had a hold on me that cannot be adequately explained by any practical attractions that fraternity membership might have had to offer. The prospect of failing to get a bid from at least one good house filled me with anxiety, for it seemed a test of my acceptability in the world of men.

With such a trial at hand, the pubescent pleasures of past Cape Cod summers—fried clams and six-packs— faded in the dog days of August. I worked the night shift at the local A&P stamping cans, slept most of the day, and grew adolescently irritable and monosyllabic. I remember laughing only once all summer, when my mother, patience drained, dumped a pan of freshly snapped string beans over my head.

Each Friday night my father drove down to our Cape Cod cottage from Boston, bringing with him a manila envelope fat with official and unofficial mail from "the institute," as I now began to call MIT. Over a period of a few weeks I collected some two dozen invitations sent from MIT's fraternities to all six hundred entering freshmen. Their variety was astonishing, and I found this in itself unnerving. From one fraternity there was a white card with tissue paper that looked like an invitation to a Wellesley Country Club wedding reception, from another a color brochure with pictures of a wisteria-draped antebellum

7   mansion, from yet another a crudely mimeographed invitation to a Dixieland beer bust. In that stack there was something for everyone, a more exotic spectrum of life options than I had known existed. In the following weeks, I had to select an appropriate identity for myself and then apply for membership in the particular tribe that occupied that niche—all to be done at a time when my adolescence was collapsing around me. I would be plagued by social ineptitude, mumbling inarticulation, unsteady eyes, infirmity of handshake.

MIT's principal residential area occupies the north bank of the Charles River. It is an odd collection of buildings strung out along a mile or so of Memorial Drive. First comes the Burton House dormitory, a conservative brick affair looking as if it had been sawed off from Harvard and rafted three miles down the Charles. Then Baker House, a 1950s-style brick housing project, wrinkled inventively into an unconventional sinuous shape, the kind of design that famous architects are supposed to create on the backs of napkins. Then a row of fraternities, including the infamous Deke house, and at the end of the row, its back turned to the others, preferring Endicott Street to Memorial Drive, the discreetly Ivy League TKE house. This was said to be the house for the prep school crowd from Middlesex and Exeter, lean blond boys with pink oxford-cloth shirts and peachy complexions who for some reason had come to MIT rather than Harvard or Williams or Wesleyan. In my last years of high school, I had developed a taste for Brooks Brothers clothes, and I selected the TKE house as my first choice.

I was a sort of demi-preppie, having attended the Roxbury Latin School, a private day school on the fringe of Boston. Roxbury Latin was déclassé on account of its being full of Jews, Irish, Syrians, and the like, but it was nonetheless part of the New England league of independent schools, and perhaps on that basis I received a hand-addressed, printed invitation to visit the TKE house during rush week. This was followed by a personal phone call at home from one of the TKE men, a Timothy Harris. "We're looking forward to chatting with you Sunday, Dick," said Timothy, speaking in a broad, full way, as if his mouth contained a golf ball.

Sunday afternoon I polished my Weejuns, put on a pair of faded but crisp chinos, a blue button-down shirt, a striped tie, and a Harris tweed sport jacket, then drove my mother's car from Newton into Boston on Storrow Drive, across the Massachusetts Avenue bridge to Cambridge. It was one of those bright, warm, New England fall days. A faint breeze wrinkled the surface of the Charles, on which the inverted Boston skyline decomposed itself into short brush strokes of titanium white, cadmium, and ochre.

On the Cambridge side of the river, Memorial Drive's traffic was light and leisurely. I drove past MIT's putty-colored limestone buildings, forlorn and formidable, down Memorial Drive, and parked on Endicott Street.

As I climbed the elegantly understated stairway, past the polished brass plate to TKE's second-floor commons room, I felt the damp tingle that comes when arteries are infused with whatever glandular juices prepare one for an encounter with a strange tribe.

The invitation was for cocktails. The affair had that subdued but elegant summery feel of August in Oyster Harbor or Larchmont. The vaulted commons room had a spotlessly Vermeeresque floor, teak-paneled walls, a backlit display of pewter mugs. A temporary bar had been established in the center of the room, from which a white-haired Negro dispersed Johnny Walker and soda, Jack Daniels and water. There were perhaps two dozen people in the room, including several women. I gazed longingly at the cameo profile of one girl, lean, blonde, face and neck tanned caramel, in fresh cotton; she leaned against the ivy-fringed casement of an open window, evidently bored, looking out over the Charles.

Harris, my host, was a sophomore, a physics major. We talked and sipped bourbon for a few minutes, then the conversation mysteriously dried and Harris's eyes began to focus in a distracted way at objects over my shoulder, like a bored date in a singles' bar. Only a year later would I come to understand Harris's position—the neophyte recruiter, uncertain of his catch, waiting to present it for the inspection of one of the senior brothers. Suddenly he locked onto his target and asked me to step over to meet one of the other fellows, a lean tennis player with white teeth named Henry Langford, whom I had seen earlier make the girl in the window smile faintly by touching her brown shoulder and whispering something in her ear. Langford was a senior, doing his degree in architecture.

"Harris told me about you, you're from the Boston area," he said in a voice that sounded as if it had been processed through a fine brass organ pipe.

"Yeah," I said, actually a sort of "yull," which was the way everyone in my family said "yeah." I mentioned something about sailing at the Wianno Club (no, I was not sure of the class, it was my friend's boat) and the name of my prep school, and about how I lived in Newton. This exhausted my meager qualifications. I was not yet experienced enough to hold my cards.

"It's a delightful city, Boston, don't you think," declared Langford, changing the subject and not pausing for my answer. "Some awfully nice buildings here; did you know that some of the railroad stations in Newton are considered to be the best examples of Richardson's early work?"

I thought of the time, in an earlier preadolescence delinquency phase, when I had broken into the Woodland Railroad Station and emptied a soda-acid fire extinguisher on the floor. Truly I was a vandal in the classical sense of the word.

"Really? I didn't know that. The Boston Public Library is really nice," I mumbled. I was sounding like a real ass. I needed something that made me an insider. "My great-grandfather worked on the library. He was disabled by a falling brick when they were building it. He was a brickmason."

Langford looked at me with a clear laser-like gaze that shriveled my spirit. "Would have been a piece of limestone more likely," he said, looking over my shoulder at someone else. "Cecil, I have to have a word with you." He shook my hand, saying it was awfully nice of me to drop by.

I watched Langford walk up to a group of people, placing his hand on one of their shoulders. The gesture stopped all conversation, and the women smiled and tilted their heads upward slightly, suspended in anticipation of his words. Currents of envy, admiration, and anger swirled through me.

I turned back to Harris. A peculiar, distracted look had settled over his face. He nervously asked me if I wanted another drink. I said sure, pronouncing it "shuah." These were the days before John Kennedy made that pronunciation acceptable.

Harris looked morose. I was feeling badly for him because I had put him in the position of having to put up with me for a few more minutes. I looked at my watch. "I guess I'd really better be going," I said.

If there was one thing they taught in those stoic classrooms at Roxbury Latin School, it was that great successes are built from the rubble of past failures. So as I drove home I took consolation in the idea that I would neither enjoy nor achieve much excellence as a WASP aristocrat. I did not want to spend my days worrying about how far to walk behind a polo pony or whether it was proper for the cook to fish for scup on the pier at our summer house.

Well, there was always the Deke house. I had visited their endless Mardi Gras twice, drank beer from the keg that was always open, seen the wild buxom girls in cutoff jeans, and rejoiced at the live Dixieland. But after a couple of visits I noticed a strained and disheveled look about the Dekes that suggested that what people said about the house was true: that the Dekes went too far but did not last too long, that their grade point average was below 3.0, the lowest on campus. Indeed I knew behind my own desperation that the Dekes were even more desperate, with good reason. Only two years before, during January's hell week, they had dropped one of their pledges, a brilliant math student (according to the tabloid *Record-American*), on the edge of a Concord woods at two in the morning, with instructions to make it back, cross-country. Two days later when the brothers returned to that same spot to try to find the lost pledge, they had little difficulty in following the boy's tracks in the knee-deep snow, out of the woods and onto the frozen lake, out to where the black hole, by then refrozen, marked the spot from which the body was later recovered with iron grappling hooks.

The Dekes, whatever their excesses—and I was not so sure, at that Kerouac stage of my life, that excesses were not virtues—at least wanted me. Or that is what a snaggle-toothed junior called Edmond told me when he presented their bid, saying I was just the kind of guy they liked at the Deke house. Edmond, whose name it seemed was always shouted, not spoken, was a jolly two-hundred pounder, a construction engineering major. Edmond liked to drink beer and displayed, both during rush week and whenever I ran into him in MIT's corridors afterward, a facial glow that I took as evidence of panicky retreat of his red blood cells from some smoking carnage in his liver. But although Edmond's bid may have prevented me from jumping into the Charles for a few days, my survival instincts, ill formed as they were, nonetheless flashed enough warnings to keep me away from the Deke house.

One morning toward the end of rush week, I dropped by

the lounge that MIT made available for commuters. The room had the ambience of a police station: walls of institutional green, rickety half-sprung furniture, the previous Sunday's *Herald* and *New York Times* scattered on collapsible bridge tables. A dusky young man with tufted eyebrows wearing a sleeveless sweatshirt sat at one of the tables. He was eating a bag lunch of stuffed grape leaves and an orange, items spread neatly in a row on a sheet of waxed paper and a brown paper bag, which, judging from their well-wrinkled condition, had been used to wrap several previous lunches. He was listening without comment to a fat boy with a crewcut and plastic-rimmed glasses who was sitting in a lime-green plush chair.

"Yep, I figure eight-oh-three is the way to go, eight-oh-three-one is too hairy for me," the fat boy said in the tone of one who has repeated himself several times and is not expecting a reply.

"Excuse me," I said, breaking the silence that followed the boy's comment. "I'm interested in finding out something about the commuters' club."

The fat boy looked at the dusky boy, who looked at me with soft brown eyes.

"Well," said the fat boy, "This is it. What is it you want to find out about it?"

"How do you become a member?" I asked.

"He wants to be a member. He wants to know how to apply for membership in the commuters' club," the fat boy said, grinning.

The dusky boy gave me a kindly look. "You just come here, that's all," he said.

I wandered over and looked at a bulletin board, which displayed several thousand thumbtack holes, and a few cards advertising hi-fi equipment and a set of drums. Then I left and drove back home to Newton. There was a message there, to call Steve at the Sigma Alpha Epsilon house. "We'd like to talk to you," said Steve; could I drop by? I got

13   back in the car and drove into the Back Bay. Inside the SAE house on Beacon Street I found Steve. Suddenly I felt the SAE house was everything I could have asked for. Was Steve going to make a bid?

"Dick, we've really had an eye on you, and all the brothers here think you're a really neat guy," began Steve.

Ninety seconds later, I was a SAE pledge.

So it was in the autumn of 1957 I became at once a college freshman and a fraternity man. The official passage occurred a few days after I accepted the pledge pin, on a Sunday. I slept late, to my father's routine disapproval, then with diplomatic caution divided the *New York Times* with him. President Eisenhower had spent the previous day in Denver, playing golf. The Atomic Energy Commission had issued a license for the first commercial nuclear reactor, at Vallecitos, California. The Boeing 707 had now run up several hundred hours of successful test flight. Bloomingdale's offered wool blankets (on sale) at $8.95. Michigan's Adrian College described its innovative orientation program, which made the incoming freshman feel a "member of the group." It featured lectures entitled, "How to Register" and "Proper Dress and Manners for the College Student," followed by a fireside sing on the beach, in which everyone participated, including the college president and the football coach. Henry Ford announced his quarter-billion measure of faith in America. The Edsel "was based on what we knew, guessed, felt, believed, suspected—about you," he said in a full page ad. And the Russians claimed that within days they would use one of their new ICBMs to fire an "artificial moon" into orbit around the earth.

That afternoon I ate dinner at home with my mother and father, and afterward they drove me in our green Buick to the SAE house. The air was crisp and full of the delicious sadness of fall, and in the slanting September sun the students arriving at MIT and Boston University, Simmons,

14   Katy Gibbs, and the nursing schools were like strokes of Renoir color against the dun stone buildings and fallen elm leaves on Beacon Street and Commonwealth Avenue. There were no parking places on the 400 block of Beacon Street, and I told my father that it would be fine if he let me off in mid-block. My parents wished me luck. I thanked them for the ride and said good-bye.

The SAE house was located in a filled estuary of the Charles River that Bostonians call the Back Bay. Originally developed as a residential annex to the civilized drumlin known as Beacon Hill, the Back Bay contained within its three square miles several buildings that I had come to recognize even as a young boy growing up in the late 1940s. There was Mrs. Eddy's Mother Church, seat of a cult as bizarre to me as Mohammedanism (in Irish-Catholic circles the news that someone's parents were Christian Scientists was conveyed in the lowered tone used to announce divorce); the squat spire of the old John Hancock building; the Boston Museum, rich with Monets and Sisleys; Symphony Hall, and across the street the yeasty warmth of my cousin Steve "Crusher" Casey's Irish saloon, where in 1957 they still danced the jig; Richardson's Trinity Church; Fenway Park, where I once became lost when my grandfather left me to shake Maurice Tobin's hand after a Red Sox game. And the Eliot Lounge, in which a tuxedoed waiter two years before had served me, a sixteen-year-old preppie (rep tie, Pall Malls, and Bass Weejuns), a bowl of popcorn and a singapore sling, thereby creating an important milestone in my adolescent odyssey toward dignity.

I entered the SAE house by climbing half a flight of stairs, opening a heavy black wood door that groaned metallically, then boomed behind me as it closed. At the top of another short flight of stairs was what my grandmother would call a parlor, which contained a piano. On warm summer nights when the bay windows were thrown open, we could sit on the curb out front and hear Ken Dobbs, an ex-sailor, playing old songs on the piano and when he got drunk, doing a pretty fair imitation of Teddy Wilson.

The chapter dining room, with its long trestle table and view of the Charles, was at the back of the first floor. Between the two rooms a staircase, wide enough to permit a gentleman to escort a properly dressed lady, led to the chapter room on the second floor. Here was the focus of fraternity social life, three black Naugahyde sofas, the magazine rack (*Engineering News-Record, Esquire, Military Engineer, Aviation Weekly*), the biggest hi-fi turntable I had ever seen, and in a small hallway next to the chapter room, the house bulletin board, a refrigerator filled with canned beer, and a slot machine. Floors three and four were study-bedrooms, and the top floor of the house was a dormitory bunkroom. A narrow staircase led to the roof, which in years past provided a platform from which bags of water could be dropped on people on the sidewalk five stories below. Now in the autumn of 1957, it was used more frequently to observe the flashing trajectory of Sputnik I.

As a new pledge, I was assigned the top tier of a steel cot in the upstairs bunkroom. Here the act of sleeping was split off from the rest of my life in a way that was new and oddly uncomfortable to me. The bunkroom had no lights and was dark any time after four o'clock in the gloomy afternoons of Boston's winter. It contained no furniture or ornamentation other than the olive-drab steel cots with their striped mattresses. The room's large windows, overlooking the wind-wrinkled brown water of the Charles, were nailed wide open, and sometimes in the morning when I woke I found wind-sculpted drifts of powdery snow on the floor. The open window convention was undisputed by any of the dozen or so brothers who slept there, and it came as a surprise to me that the consensus should call for winter sleeping temperatures at or below freezing; I wondered on this account whether the rest of the American people were hardier than those of us from Boston might imagine. It was here too that I learned that many young men sleep in the nude, an observation that created in me, a life-long pajama wearer, an oddly uncomfortable resonance. And the ease with which others fell asleep! To be free of elaborate rituals, lowered lights, bedtime reading, the Lone Ranger; rather to walk briskly up the narrow staircase into that black chamber, shuck slippers and robe in the teeth of an icy wind moaning off the Charles, leap into bed, and within minutes, release one's grip on precious consciousness. It all seemed so healthy, uncomplicated, and, well, uncivilized.

Three of us shared a study room on the third floor. It had capped gaslight fixtures, and wood trim swollen with multiple coats of paint. Dust balls danced beneath an eclectic collection of used furniture whenever anyone opened the door. From our grimy window we had a fine view across the Charles of MIT's gray limestone buildings spread like a Mayan city on the Cambridge plain.

David Wilson, a fellow pledge, had the least desirable desk, next to the malodorous closet. Wilson was an eager boy with a prickly black crewcut and nervously explosive laugh. He had graduated from high school in Orlando, Florida. A repulsive exhibit featuring preserved pieces of a cow's eyeball had won for him first prize in the state science fair. He wanted to be a scientist, of course. But by the following May, his inability to grasp the writhing abstractions of differential calculus caused his grade-point average to sink below 2.5 for the last time. I imagine Wilson a prosperous optician in some Florida shopping center, thanking his stars that he left Boston at the end of his freshman year.

Cluttered with books and papers and spilled ashtrays,

the other corner of the room contained a bed, in which until noon on most days the semiconscious body of Percy E. Lee, Jr., could be found wrapped in a depressing tangle of gritty sheets and ulcerated blankets. Percy was a kindly young man from Palestine, Texas. A junior, he was assigned as my big brother, and I got to know and like him in the first weeks I lived there. I never understood what fatal misconception brought Percy to MIT; certainly he was no engineer, and during the two years that I knew him I watched his spirit scorch and wither, yet nonetheless endure, in the thin dry atmosphere of the institute. Of the three of us, he was the least suited for fraternity life and as a pledge two years before had been the target of that prurient bullying the weak enjoy inflicting on the vulnerable under the institutionally approved guise of building character. Percy's unceasing inner struggle with such basic concerns as the condition of his internal organs and the existence of God gave him a distracted, preoccupied manner, and when he talked his eyeglasses, which were usually temporarily repaired with Band-Aids or paper clips, slid down the bridge of his nose.

Percy talked in a bucolic southern accent with an intense and usually inappropriate sincerity that created a sense of mystification and anxiety in most of his listeners. He had no interest whatsoever in his major, chemical engineering. He was, however, greatly interested in comparative religion, and he sometimes pressed me to read philosophical tracts by Tillich and Barth, which I failed to understand. I think that Percy aspired to be my mentor, and I have thought since then he would have made a fine professor. I remember once asking him to explain metaphysics, which I was straining to think of as a branch of physics.

"Well, if you take this heah quarter," said Percy, holding up a silver one for me to see, "and fuhst you take away its mass, then you take away its visibility, an everythin' else you can observe about it, an you think they's still somethin theah, well that's metaphysics."

Four years at MIT permanently solders some primary circuits of the mind, and perhaps for that reason certain modes of thinking seem inalterably closed to me. A few years ago I tried to read an elementary interpretation of the metaphysics of Saint Thomas Aquinas, said to be easily grasped by college freshmen at Georgetown University. I understood nothing of it. The blind spot persists; only last night while conversing over a bottle of Côte du Rhone with a companionable Armenian priest, I encountered the same mysterious voids in my comprehension. Only Percy's definition remains, a pinhole of light.

Percy eventually graduated from MIT, switching majors on the way to course 15, "Industrial Management." To Percy engineering had been merely pointless, or absurd, to use a word he favored. Industrial management, on the other hand, had the added dimension of being evil, and his hatred of the subject matter seemed to provide him with the necessary energy required to complete his degree in it.

I lost track of Percy for several years while I was in the army and, later, living in the Far East. I heard that he was working as a computer programmer at Southern Methodist University. This was no surprise; that particular trade provides a refuge for a significant percentage of my generation's lost, violent souls. Is not FORTRAN technology's Latin, the programming of computers a form of manuscript copying and illumination?

Once when I had returned to the States after a three-year absence, I visited my parents on Cape Cod one July weekend and was surprised to receive a call on Saturday afternoon from Percy. He was visiting his new in-laws only a few miles away in Falmouth, he said. Could I make it for dinner? I said I would and drove to Falmouth that afternoon. The in-laws' house was a two-story summer cottage, gray shingled with a screened porch, jammed into a small lot several blocks from the beach, in what easterners call a summer colony. The yard was crowded with cars and family and friends. Everyone was feasting with merry rapacity on barbecued beef ribs, A&P crinkle-cut potato chips, and Table Talk pies, and there was a great deal of champagne being passed around in mugs and fruit jar glasses. Percy greeted me by shouting my name and introduced me to his new father-in-law, Harry. Harry was involved with concrete, a subject that provided some common conversational ground, for I had worked in heavy construction for the past two years. We chatted briefly, and then Harry went off to tend the fire.

Percy was a little drunk, the eyeglasses were low on his nose, and soon he began to deliver a diatribe in which Lyndon Johnson emerged hazily as the target. I fell silent after concluding that whatever fragments of my own recent history I might volunteer would too easily be thrown into the cesspool of American values that Percy was declaiming with such passionate intensity. Yes, he was still programming computers, he told me when I changed the subject. His work was a joke, just absurd. "You see, Dick, I've really arrived, it's the American way," he said with self-deprecating glee, raising his voice to a shout on the last two words. His wife was looking at us, frowning, and conversation around us began to die. It was as if Percy, in gathering force for some terrible question or pronouncement, was magically draining the conversational energy from those around us.

But the father-in-law sensed the danger and walked up to us, placing a hand on Percy's shoulder.

"How you fellas doin; you got enough to eat there, Dick?" he asked me.

"Right, I'm fine," I said. I was holding uneaten an enormous beef rib, gleaming with grease, on a small and structurally inadequate paper plate. Percy wandered off to refill his mug with more champagne. Harry looked at me dubiously.

"Percy's actually a nice guy," I said, immediately wishing I had left out that didactic "actually," with its double implication of superior knowledge and desire to be helpful. I do not like being on the receiving end of that, and I certainly had enough experience to know that no one mixes those two messages together in the construction business. But Harry let that blunder go with a long, unfriendly look and then left me with the suggestion that I help myself to more food.

Percy came back. He had passed on to a new, more docile plateau of intoxication, and now he seemed bored with my company. I stayed a few more minutes, then excused myself and left, driving home on back roads, through that forlorn pygmy landscape that lies a mile inland from Cape Cod's crowded summertime coast. It began to rain, the saddest rain I have ever seen. As I drove, I wondered why Percy so compulsively put himself into situations in which what he had to offer—that earnest explanation of metaphysics, I supposed it was—meant nothing to anyone around him. I had turned twenty-eight the month before, and I knew that so far I was not doing much better with my own life. It would be a while before I came to understand that many of us spend that third decade of our lives fleeing, in fear and loathing, from what we are.

Nineteen fifty-seven was a good year for the SAE house. Rush week had yielded a full crop of seventeen pledges. We were formally invested in a candlelit ceremony one evening the first week of the term. On each of us they pinned a gold and blue enamel pledge pin. We heard a grave ceremonial speech by Chuck Wilson, a senior electrical engineering major and elected leader (eminent archon) of the house. Afterward all of us, brothers and

21   pledges, held hands uncomfortably in a great circle and
sang songs. This event, like most other ceremonial functions in the SAE house, was directed by Alan Hart, a
twenty-five-year-old graduate student of fuzzy academic
status, our unofficial priest and repository of tribal tradition and lore. Hart was a starter of songs; any brother who
brought a date to dinner could count on Hart's clinking
knife on glass at coffee time and announcing with a gravity
that did not entirely conceal his eager sentimentality, the
sacred word *violets*. Then we would all be obliged to sing,

Violets, violets, you're the fairest flower to me
Violets, violets, emblem of fraternity . . .

Matters were different when there were no female guests
present. Even now as I write a continent away and nearly
twenty years later, I feel the adrenal twitch. There was the
dreaded clink of stainless steel on glass, and the eyes of
the brothers were hard upon me as I rose to fumble
through various fraternal mysteries, names of founders,
the belle who hid the charter during the Civil War (Miss
Lucie who?).

"Identify this brother," I was told, and I peered across
the candlelit table at Dick Baker, who froze self-con-
sciously. Baker was said to be a brilliant physicist, nearly
straight As all the way through his undergraduate years.
Now he had an office in one of those old army barrack
buildings in which MIT fifteen years before had perfected
radar. He led our review sessions in freshman physics, and
his enthusiastic affection for abstractions as Maxwell's
equations and the Poynting vector provided me with a
Dantéesque glimpse of that underworld of currents, fields,
and particles.

"Richard William Baker, course eight," I said, then
paused while my mind scrambled over that childhood
map that I still carry in my head, past yellow Ohio with its
picture of a tire, green Indiana with ears of corn, to pink

California with its roll of movie film. "San Marino, California. He's engaged to Susan Crittendon."

"WRONG, PLEDGE," shouted Tom Maguire. Maguire was a bastard, a junior, an Edmund for King Lear if I ever saw one.

"Try that again," says Baker, pleasantly, without the simulated outrage of a sophomore or junior.

I blundered through a list of Lucys, Cheryls, and Sallys until I was dismissed as inadequate. I liked Baker and his girlfriend; I guess it was Sally. But I wondered how Baker's mind could have such a terrible flaw that he had never been able to pass any of the "Foundations of Western Civilization" courses that MIT requires of its freshmen and sophomores. Surely a sensibility that finds Rayleigh-Jeans law no more complicated than a traffic regulation should be able to write an acceptable paragraph or two on Voltaire or Lenin. But Baker's mind was evidently shocked into fibrillation at the mere mention of Western civilization, and he had never been able to pass the required courses. Hence one of MIT's hottest physicists held no college degree. I've since observed that it is the people like Baker, driven and purified by secret inadequacies, who become the guiding geniuses (but not majority stockholders) in companies on Route 128 and the Santa Clara Valley, that produce miniprocessors and intelligent silicon chips. It has always seemed odd to me that a mind capable of harnessing for practical purposes the abstraction of a collapsing magnetic field balked at Plato's shadows on the wall of a cave.

Were those pains last month an innocuous spasm of the pylorus or a cardiac infarction? Perhaps I should hedge my bet, as Descartes did when he decided to humor God just in case there was one. My personality may be ruining my arteries. I am always late to the airport, especially for flights to Houston. A famous cardiologist suggests that we

23   should enrich our days with more ceremony. It seems that
we have stripped ritual from our lives, dosing ourselves
with Valium to treat ills more effectively cured with whiffs
of incense. Consider the Masai. They eat enough cholesterol every week to lubricate a 1957 Oldsmobile. What is
their secret of long life? I submit it is this: when the Masai
girls go off into the bush with their friends to giggle about
female matters, the boys are wisely taken aside by the
tribal elders and taught the secret of the bullroarer, a mere
stick that when twirled on a string makes a noise frightening to women and children. In this way they become
men. But in Framingham, Massachusetts, adult males
never get it straight in their minds whether they are men
or boys. This is very stressful and, in ways that are not yet
clearly understood, causes clumps of cholesterol to be released into their arteries with devastating effects.

Wisdom is a matter of patient archaeological excavation;
each week of our past is a sort of geological stratum of
minutes, meals, changes of underwear, heartbeats, a clut-
ter of artifacts in various stages of preservation. The living
flesh has been consumed by pathogenic agents, long since
leached away, but the lessons to be learned are preserved
in the almost accidental position of flint and bone. I excavate through the dusty layers of four years at MIT until I
strike January of 1958. It is a rich midden.

Hell week at SAE was scheduled immediately following
Christmas vacation, a time when the MIT freshman's attention might be temporarily diverted from the indecisive
valences of chromium, the diagnostic utterance of the bal-
listic pendulum, the plucky recovery of the Naperian loga-
rithm from the trauma of integration.

The burlap shirts come to mind first, the coarse scratching against the nipples, which grew red and swollen by
the end of the week. Procurement of the shirts had been
my first task, that Sunday night when the brothers, howl-

ing and red-faced, pulled us from our beds and sent us out on strange quests. I was to return by dawn with seventeen potato sacks. Dave Searle and I went out together, taking the MTA from Back Bay to Park Street, then walking down to Haymarket Square. The night was icy clear and still, stars dusted the blue-black sky. We found an Italian unloading produce from the back of a truck and told him that we needed potato sacks. He took us inside the warehouse to his brother who sold us seventeen of them, in good condition, for five dollars. The Italian brothers did not seem to think it strange that two young men dressed in chinos, desert boots, and loden coats needed burlap sacks at three o'clock in the morning, and they seemed to know there was a certain importance in it. They asked no questions and carried out the transaction with a certain gravity. Searle had other matters to attend to: five dozen fresh eggs and one hundred feet of rope, which we found easily enough. Then we walked back to Park Street station.

Inside the station the air was warm and moist, lightly perfumed with ozone and urine and Boston blue clay. We waited for a car, talking of our favorite books, of Daisy Buchanan, and Mac and the boys, of packing freshly caught trout in ferns on hot days. Then after twenty minutes or so an orange MTA car marked Cleveland Circle came rocking and screeching down the tunnel. We rode to Massachusetts Avenue and then drank coffee at Hayes-Bickfords. In a lavender dawn we walked back to the SAE house.

That night, before disassembling and cleaning the house stove, we tailored our sacks by cutting out holes for our heads and arms. Then before the assembled brotherhood we put them on as undershirts, a ribald investiture.

The official purpose of hell week was a secret, revealed only in the catharsis of its consummation, and protection of this secret required that we pledges be maintained in a sleepless condition throughout the week. During the days,

25   we were required to attend all of our classes, itching in our potato sacks, which we wore at all times beneath our shirts. Back at the house in the evenings, we were harassed at the dinner table and then assigned to various scrubbing, cleaning, and fixing tasks, which were interrupted from time to time by ceremonies in which we were forced to eat raw eggs, behave like barnyard animals, and recite self-deprecating rhymes. Late in the evening, we were ordered to bed. Then after what seemed to be minutes, the brothers fell upon us, shouting and beating pans, and we were once more set to work. In this way I got very little of that deeper sleep that knits together the easily unraveled fabric of my mind. By Friday, five days into hell week, dreams, memories, and reality were flashing on the white screen of my consciousness like a collection of thoroughly shuffled photographic slides.

Sunday night, we had been warned repeatedly, was the climax of our trials, the night of the examiner. The examiner would test our worthiness for lifelong brotherhood, and for subsequent membership in the chapter eternal, as the organization of dead SAEs was called. It was rumored that the examiner had little patience with MIT students, whom he suspected of being more interested in angular momentum and directional derivatives than in the middle names and birthdates of past grand archons, those balding hardware manufacturers and insurance agents whose re-touched photographs filled the hardbound volume of SAE's official history.

Then Friday evening something terrible happened. A meeting was suddenly called in the chapter room. Chuck Wilson, our eminent archon, stood before us all, pledges and brothers. He wore a worried frown. There had been a mixup, a disaster, he said. The examiner had arrived two days early; he had called and said that the examinations would have to take place that very evening. No amount of urgent pleading had been effective; the examiner was brusque and impatient; he did not care whether the pledges had not slept all week, that they had not yet been instructed in important details of fraternity history; these MIT pledges were supposed to be brilliant, weren't they? We listened to this news numbly. The brothers gasped and murmured anxiously among themselves. That was the end of hell week, right there, Chuck said. We should shower and change and review our material. "Just do the best you can, you guys," he said.

At seven o'clock in the evening I reported, scrubbed, clean shaven, and at last free of my burlap undershirt, to Dick Baker's room, where Baker and Ed Blake were to assist me in reviewing my material while I waited my turn to be examined. Both of them were nervous and distracted, and Baker kept scurrying downstairs to assist in providing appropriate trappings for the ceremonial examination, which was to be held in the Ping-Pong room in the basement. The examiner had arrived and was said to be impatient and irritable. "I don't know about this guy," said Blake to Baker, who was trying to impress upon me the name of the mansion where Miss Lucie Masters hid the SAE charter during the Civil War. I was making little enough headway at this; when Baker asked me Miss Lucie's date of birth for the third time, and I answered 1939, my own birthdate, they decided that I should rest for a spell.

The room grew quiet. A radiator ticked, and occasionally I heard footsteps and muffled voices on the stairs outside. Baker sat in shadow at his desk, his face gleaming in the glow of his desk lamp. In his peripatetic way, Baker had recently taken an interest in information theory, and now he was reviewing some experimental work having to do with the capability of subjects who were seated in a perfectly dark and silent foam-padded room to discriminate patterns among sequences of numbers heard through a set of earphones. The work was one of those odd brews of mathematics and psychology that was very close to the core of MIT's new world view in the late 1950s.

For perhaps an hour or two, I hovered in that half-world in which reality becomes miraculously plastic and admixed with visions, and from time to time Baker and I were joined in the room by a motley crowd of well-wishers, including my sister, a former camp counselor, the cranelike proprietor of a local drugstore, and the eminent mathematician and child prodigy, Norbert Weiner.

At about ten o'clock two of the brothers entered the room. One of them produced a blindfold, which was tied around my head. I was led from the room, down two flights of stairs to the dining room. Past the pay telephone (which could be activated free of charge by jiggling a pin into a hole in its lead wires). Past the slotted wooden rack in which the daily mail, peach-colored letters from Duke and SMU announcing that one's high school sweetheart had met someone else, terse greetings from Dad dictated and typed on company letterhead and enclosing a check for $150, flyers from national headquarters in Evanston announcing the availability of official SAE boxes for storing past issues of the fraternity magazine. Down the narrow flight of servant-width stairs, to the basement of the house. Now you just relax, Dick boy, we're all behind you tonight.

We enter the Ping-Pong room. A different feel to it tonight, like a crowded church, whispers and burning wax and warm wool, coughing, shifting of bodies. Just kneel down, Dick. Anticipatory silence. Bless me father, for I have sinned. Gentle hands untie the blindfold. I blink.

The examiner's face is illuminated by the flames of two candles that stand in tarnished silver candlesticks on a felt-covered table. Chuck Wilson and Tony Biggs, his subarchon, sit in darkness on either side. The examiner's face is four feet away but it seems to grow until it fills my field of vision. The eyes are gray, expressionless, the skin and lips are waxy, almost translucent. The face turns to Wilson.

"Is this the last one, or what?"

"No, there are two more," says Wilson.

The examiner looks annoyed. His voice is dry, with a hard edge.

"Well, let's get going here," he says. "O.K. Sheehee, have you studied this book; do you know the history of this fraternity?

"Meehan," says Wilson. "His name is Meehan."

"All right, Meehan," says the examiner. Then he asks me some questions, some easy ones, which I answer, then the year of the typhoid epidemic at Washington and Lee that did away with Chapter Zeta, which I do not know, then one that I get, then two that I do not get.

"I thought you said you knew this material."

"I do, I mean I did." My voice sounds small and shaky, and I cannot breathe deeply.

"Look," says Wilson in a low voice, "he's tired, give him a chance; he knows it all right."

"I'll decide that," says the examiner sharply.

The crowd behind me shifts and mutters.

An easy question, a blooper, which I get. Then I miss three fast balls in a row.

"You call yourself prepared for this examination?" the examiner says. He turns to Wilson. "This won't cut it, gentlemen, I can't pass this man." Protests from the crowd. It's not fair. Give him a chance.

"We've got standards here, and this man doesn't meet them," says the examiner. "That's all. You can bring him back next year. Let's have the next man." Ripples of outrage in the crowd. The blindfold is refastened, and I am led out, up the stairs, back to Baker's room.

For an hour or so while I lie on Baker's bed, there are desperate negotiations. Various brothers pop in and out of the room. This has never happened before. Chuck is really upset; he's trying to talk the examiner into a retest. I'm the only one who has ever failed.

I look at the wall. Baker has pinned a cartoon on it. It shows a smiling idiot, with a slide rule. "Four years ago I couldn't even spell engineer, now I are one," reads the caption. The brothers are doing the best they can for me, miserable failure that I am. All I ever wanted was to be a fraternity man. My head hurts, my diaphragm has been locked in a spasm for hours, my eyes fill with hot tears.

The pinning ceremony took place in the chapter room a few minutes later. Thirteen pledges stood in a row before the assembled brotherhood. Our blindfolds were removed. The examiner stood before us and declared in a solemn voice that the results had not been what he expected from MIT men, but that they had been generally adequate. That was, with the exception of one man who had not met the standards and could not, he was sorry to say, be admitted to the brotherhood. Perhaps next year. That man knew who he was and would have to be excluded from the following pinning ceremony. That man should now step forward.

Thirteen of us stepped forward.

It was a joyous evening, with drinking and shouting and singing songs of brotherhood, and I can remember myself saying over and over again how great it all was, how I had been totally fooled!

It is only since moving away from the East that I have come to understand truly that a spell of bad weather occupies but a small part of the world and that it always passes in time. I have since seen through the eye of the satellite how the whorled perturbation of a storm can be framed within a degree or two of latitude and longitude and have flown in a tossing little Cessna around and even above the squalls of a front.

The storms of my youth in Boston were different, or so
they seemed to me then, when I would awake in a universe
of blown snow, to the white noise of its fall. I thought then
in the basement of my mind that the same snow was falling
on Battle Creek, Michigan, and China and Tierra del
Fuego, and that it would continue to fall forever, or almost
forever, all the day long and throughout the next black
night.

So it felt to me when the first big snowstorm of the season came in February. I had a calculus midterm that morning, and I arose shivering in the gray light of the
bunkroom. The floor was powdered with snow; I made
dark, wet footprints on it, then stumbled down the musty
warm staircase to my study room. Wilson had been at his
desk for two hours, huddled in the glow of a lamp, bent
over his gray and red Thomas's *Calculus and Analytic Geometry*. A novitiate at his breviary:

Let $y = f(x)$ be continuous for $a \leqslant x \leqslant b$, and possess a deriv-
ative at each x for $a < x < b$. Then there is at least one number
c between a and b such that $f(b) - f(a) = f'(c)(b-a)$

Unfortunately for Wilson the calculus examinations at
MIT gave scant attention to such formalities; rather they
presented for solution what we engineers think of as more
practical problems. These were often of the professor's
original design and had to do with the relative and absolute spatial location of a mythical MIT student. The student would be presented as walking his dog on a leash,
when some mathematically defined catastrophe—the dog
chasing a cat around a lamp post, dragging him after it, for
example—shattered the Cartesian linearity of their movement, introducing an urgent need for polar coordinates or
directional derivatives. Wilson thought such questions
unjust, for there was nothing in Thomas's book about men
or dogs.

I switched on the converted gas fixture lights to find my

clothes. Percy, who had doubtless been up studying until dawn, muttered a complaint and shifted about in his corner bed, so I turned off the light, putting on a pair of army surplus wool pants and a wrinkled Viyella shirt in semi-darkness.

Downstairs the house was quiet; the nocturnal pianist Ken Dobbs sat alone at the dining room table drinking coffee. I joined him there, ordered breakfast from Joseph, our house waiter, and flipped through the remnants of Sunday's paper. Vanguard had fizzled on the pad while Sputnik beeped its A-flat tone serenely overhead. The Japanese were tactful but disappointed, the Polish army newspaper considered the much-publicized Vanguard fiasco a "moment of merriment in the dull drabness of everyday life" and the *New York World Telegram* editorialized that "the sound we want to hear is the thud-thud of heads being knocked together in Washington." A Sunday supplement carried an article describing the predictions of eminent scientists for the year 2057. Werner Von Braun visualized colonization of the moon, on which "lavish excursion hotels have been established; they run the gamut from honeymoon hotels to wide open gambling joints." A Cal Tech biologist predicted that we would all be vegetarians. "We will eat steaks made from extracted vegetable protein, flavored with tasty synthetics, and made chewy by the addition of suitable plastic matrix," he said.

At twenty minutes to nine I finished my coffee and eggs, put on my gray wool coat, and buckled on black rubber boots. Then I went outside, the door booming shut behind me. The dry snow brushed my face and squeaked underfoot as I walked the block to the intersection of Massachusetts Avenue and Beacon Street. There I stood alone for a moment looking back at my footprints coming from the SAE house. So I had at last been shown the bullroarer. But now the excitement had died down, and already some of the new brothers were talking about next year's pledge class and hell week. I could find nothing to look forward to in that. But thinking about the moment a month before, when I had stepped forward, alone, I remembered how there had been a certain incomparable exhilaration in that instant. I had failed, and yet I was free. I knew then that I had come of age in my own way and that I was free now to leave the brotherhood.

The facade of the Mount Vernon Congregational Church and bare branches of the street trees were a soft umber aquatint on grayish-white paper, and across the hushed street someone switched on a yellow light in one of the apartments on the fourth floor. A stewardess, perhaps, wrapped in a warm robe, voice soft and husky with sleep. A car approached, chains clanking, cutting fresh tracks in the snow. I held out my thumb, displaying my notebook and hunching in a way that seemed forlorn and scholarly. The car stopped, and I got into the front seat and exchanged greetings with the driver. Then we drove over the bridge in the snow, across the Charles to Building Seven.

Not far from my California home is an obscure library that contains among its curious collections a set of catalogs of MIT for the period 1910 to 1977. The volumes of this collection comprise a sort of upturned geologic stratum of techological philosophy, from which I have extracted three specimens for examination.

The first of these, which I have open before me now, is the MIT bulletin for academic year 1957–1958, my freshman year. It is a hefty tome of 372 pages with a pleasant musty smell and sepia photographs of fluted limestone columns, front and back. It contains a good deal of discussion of the aims and directions of mid-century American technology, as well as many interesting photographs. In it I read that it was at MIT that the disciplines of chemical, electrical, and aeronautical engineering originated and that the institute's curriculum is aimed at producing something more than narrow-minded technicians. Its undergraduate programs "are so broad and fundamental as to constitute an excellent general preparation for careers in other fields," according to James Killian, president, and later Eisenhower administration science adviser. Dr. Killian's educational philosophy is to maintain "not only exacting academic programs directed to a high level of intellectual achievement but also to prepare for life and citizenship." MIT is "a university in the truest sense—a university with selected objectives, built around science and social technology but embracing the arts, the social sciences, and the humanities as essential partners." Many of the catalog photographs feature serious men tinkering with various engines, among them a proton cyclotron, a

34   subcritical graphite-moderated pile, the Rockefeller Van de Graaf generator, a ship modeling table. One photo shows a serious young woman adjusting a stopcock in the biology laboratory. There are pictures of square dancing, the new Kresge auditorium and chapel, and Eleanor Roosevelt, who is delivering a talk sponsored by the Lecture Series Committee. The most popular photographic theme is a group of young men, hair trimmed short around the ears, heads thrust forward alertly, contemplating a graph or equation, a billiard ball or a dish of moldy jelly that is being described or manipulated by a pleasant older man wearing a stringy necktie.

For those of us who are to take up careers in engineering, the catalog explains that the objective of our education will be "to give expression, direction, and discipline to the urge, inherent in so many of us, to design and build, and to participate in man's efforts to harness the forces of nature for his benefit." Yet we will have to be conscious of the social and economic implications of our decisions, and our real stature will be attained in those situations "where judgment and wisdom shape the elements of scientific knowledge to achieve results which will benefit our society." I understand the part about harnessing nature, but I am not sure about that last sentence, although I have read it several times. Whose judgment and wisdom? The technologist's? What is "judgment" anyway? Are we scientists and engineers supposed to decide what "results" are to be achieved? Is scientific knowledge a kind of pure raw material that can be "shaped" to achieve results (as most scientists claim and perhaps believe) or is it already "preshaped" at inception for some intended use (as Francis Bacon said)?

A few pages further on, civil engineering is described as concerned principally with the "betterment of modern civilization." The civil engineer "changes the geography of the earth," producing an "improved environment." The

35   freshman elective civil engineering course is "Technology in Our Civilization": "Fundamental resources of civilization: power, materials, and man. Principal technological agencies of civilization: transportation, communications, environmental control, food production."

All freshman engineering majors will concentrate on fundamentals: 70 percent of their forty-eight-hour week will be spent on calculus, physics, and chemistry. Required in addition will be a general Western culture humanities course, a course of military science, plus an elective. Possibly "Man's Food," or "Basic Machine Drawing," or "Public Speaking."

Dr. Killian had been a student at MIT three decades before me. His senior year there, 1926, had been a banner year for technology. In nearby Auburn, Massachusetts, Robert Goddard had sent a liquid-fueled rocket soaring forty-one feet into the sky. In Berwick, Pennsylvania, a new automatic potato peeler had increased Earl Wise's production so much that he moved his potato chip factory from his garage to a 32' × 75' plant. It was the year that the Model T began to lose ground to Chevrolet; of the first scheduled airline service; of the Santa Fe Chief. A year of American invention, of waterproof cellophane, synthetic rubber, talking movies, and the Garand M-1, of "The Little Engine That Could."

The MIT catalog of that year, 1926, is a smaller, yellowing, mustier book. Less rhetoric and philosophy, more lists and tables, no photographs. It advises that the purpose of the institute is to afford to students training as will fit them to take leading positions as engineers, scientific experts, teachers, and investigators. MIT's location in the neighborhood of a great manufacturing district is of great advantage. It is assumed that students come to the institute for a serious purpose and that they will cheerfully conform to regulations. Tuition is three hundred dollars per year. The

36   Samuel Martin Weston Scholarship Fund offers aid to a native-born American Protestant boy, preferably one from Roxbury. My field, civil engineering, is described in practical, concrete terms. Surveys. Construction. A hydroelectric specialty option. Docks. Waterworks. Training in a compact body of principles.

The 1926 freshman year requires fifty hours of exercises and preparation, 62 percent spent on basics—math, chemistry, and physics—a lesser percentage than in 1956. English and history, which is limited to European history since the dawn of the industrial revolution, is required, as are descriptive geometry, mechanical drawing, and military science. In comparison to 1956, the emphasis is relatively practical, technical. Civil engineers are required to take a course in astronomy and spend a summer at Camp Technology. Perhaps it was there that the student was trained "to have courage and self-reliance in solving the problems that the engineer has to meet".

I've never met Dr. Killian, but I am no stranger to Dr. Killian's class of 1926, for my father was also a member of it.

I pick up the telephone to call him. His voice comes over the line, three thousand miles and a generation away.

"What's the first thing that comes to your mind about your freshman year at MIT? Do you remember your courses?"

"Sure. Calculus, chemistry, and physics. Wait a while. What was that other course? I forget the . . . "

"Descriptive geometry?" I have the 1926 catalog open before me. "That's it," my father says. "That was a bastard. It was a strange course. They'd say something like 'draw the intersection of a cylinder and a cone with axes of such and such.'"

I ask him what it was like, that year. The mortality was very bad, he says. Almost 50 percent of the first year. Socially, it was pretty rough. A hell of a work load.

I ask him how it was different then than now. Well, take the matter of structural details, for one thing, he says. Work today is sloppy. No one knows how to detail. Worse, no one seems to care. Last week he had been looking at an inverted truss design. The last web member was displaced to clear a column. "The result was the last web member coming up to the chord was eccentric with respect to the column. There was a *moment* at that support. The bending moment produced a combined stress of 73,000 psi for 50,000 psi yield point steel. I pointed it out to the designers. You know what they wanted to do? Tack a plate on there. Without unloading the truss."

I visualize my father shrugging at this, as if to say "that's the way things are done these days, it's not my way, but what can I say?"

I thumb through my most recent specimen, the 1975–1976 catalog. What will come of those molded by its philosophy? The graphics are sexy, youthful, and candid—T-shirts and bicycles and cute girls—as if it were to be read while listening to Simon and Garfunkel. President Wiesner's opening statement is a defense of the social role of science and technology. He wants MIT to lead the way in relating "our humanistic needs to the choices available in technology, social organization, and ecology." The catalog reads as if the faculty and administration have just spent a sabbatical year at Williams or Brandeis. MIT's formerly battleship-gray walls are repainted in pastels and earth tones. Interdisciplinary and experimental programs sprout like ivy from every limestone crack: "Crossroads in the Western Tradition." "Archeology and Ancient Technology." A definition of civil engineering: "The essence of the profession, as we view it, is to bring about a symbiosis of the constructed facility with the natural environment on the one hand, and with the social environment on the other."

In half a century, action has yielded to contemplation; analysis to anxiety; compulsory chemistry to optional biology; the list to the matrix; map reading and topographical drawing to decision theories and social realities in engineering planning. I recognize in the catalog the name of a friend, a research assistant in 1976, now a research program manager in Washington. I call him.

"I can hardly believe this. The greening of MIT." I am impressed with the changes in the catalog.

"Oh it's all gone back again," he says. "They got rid of the anthropologists and women. A counterreformation. The Roundheads are back."

A damp box from my garage yields material pertinent to my retrospection: a bulging file containing, among other documents, an audiological report of near-deafness in the 3000–6000 Hz range; an abrogated trust agreement with a woman to whom I am no longer married; a certificate of health for permanent settlers of Brazil, filled out in a fit of escapist desperation but never sent; a Pakistani exit visa (No. 9262 via Torkham Khyber, August 22, 1965); honorable discharge papers from the U.S. Army. Among the lower strata is a photostatic copy of a grade transcript, which indicates that MIT student number 570497 completed, in what must be described as a mediocre performance, the following standard freshman courses in the year 1957–1958: general chemistry, military science, physics, calculus, foundations of Western civilization, and philosophy and scientific methodology.

Of general chemistry, little remains beside the delightfully sudden blush of phenolthalein and the sharp pungency of the student laboratory, pleasant because it disguised something worse, a certain factory stink (mercaptans, our instructor said) that permeated Cambridge on overcast winter days.

Of military science, the impressions are more vivid. On

Wednesdays, drill days, I wore a Korean War officers' uniform to class. In winter it was already dark as I dodged traffic on Massachusetts Avenue, rushing to the sooty brick armory, slush seeping through the thin soles of army-issue shoes. In spring it always seemed to rain on Wednesdays, slanted rain, so the upwind side of my blouse and trousers became limp and baggy.

We picked up our M1s from the racks in the basement of the armory, then went upstairs to a basketball court with creaking varnished floors and echoes. There we practiced the manual of arms for an hour. Although ROTC is said to have originated at MIT, not many of us were traditional officer material; our COLUMN LEF, HARCH! came out in little barks rather than with the fully rounded roar heard at a place like Boston College.

I hated drill, of course, believing it was a waste of time. Now I look back on it from the higher ground of middle age, a vantage point from which it is not so clear what is and is not a waste of time. My view of drill has softened. It seems now, in retrospect, that it was relaxing, a form of meditation. The soldier has few responsibilities while marching; it is a pleasant return to a world where essence precedes existence. I was a soldier, so I marched. It is a relief not to feel so important. I tried to explain this, and how the draft really was not so bad after all, to a class of Stanford graduate students last year. They were dubious. Some of them looked at me as if I were mad.

The guts of one's freshman engineering education were calculus and physics. Physics begins with Lord Kelvin's dictum, "When you can measure what you are speaking about, you know something about it; but when you cannot express it in numbers, your knowledge is a meagre and unsatisfactory kind." We began by playing with toy cars that ran on tracks, bumping into each other and running down ramps. After a while we learned that the cars and the tracks could be replaced with symbols m's and x's and y's, and soon enough we could play with those abstract symbols instead of the cars themselves. Calculus made the symbols behave like the cars when gravity got into the act. I was impressed to learn that most of these matters were discovered by Isaac Newton during eighteen months while his school was closed due to the plague. It took me as much time to learn these tricks as it took Newton to invent them.

The required humanities course, "Foundations of Western Civilization," concerned itself mainly with history and philosophy of classical Greece and included readings from Plato, Thucydides, and the dramatists. According to President Killian, the humanities program was supposed to "deepen students' understanding of themselves and their environment".

My elective course, "Philosophy and Scientific Methodology," was based on a textbook entitled *An Introduction to Philosophical Analysis*, by John Hospers. This course focused with almost benumbing thoroughness on such matters as whether trees make sounds when they fall in forests, and it left me with the queasy idea that if one doggedly chopped away at any conventional wisdom, one would find very little truth worth preserving in the end. This school of inquiry, logical positivism, was intended to clarify linguistic matters. It provided little consolation for tender-minded freshmen seeking answers to the problems of life. It has reportedly fallen out of fashion, given way to structuralist trends.

This is not to say that my introduction to philosophical analysis has not been of some subsequent use. For that matter, the author of our text, John Hospers, later sprang from his professorial armchair to become an early theorist and presidential candidate of the Libertarian party, the platform of which might be considered the result of application of the positivist razor to conventional political programs. Libertarian doctrines have enjoyed more than modest success these days; Hospers should be pleased to have synthesized an influential ideology from what might have been considered a dry and pointless academic exercise.

During those same months that those of us of the class of 1961 were becoming acquainted with Socrates and Newton, an exciting technological event was taking place three thousand miles away on a cattle ranch in Vallecitos, California, an hour's drive east of San Francisco. The nation's first commercially licensed nuclear power reactor was, in the unfortunate terminology of the nuclear physicists, "going critical," starting up for the first time. Built jointly by the Pacific Gas and Electric Company and the General Electric Company, the Vallecitos nuclear power plant was a puny contraption by any present standards; its turbines had been confiscated from a Russian ship that had blundered into California waters. Nonetheless there were those who saw in this primitive engine the kickoff to a boundless age of technological progress. "A world-wide golden age is truly within our reach," said Harrison Brown that fall, predicting a long-term shift to nuclear power. And "progress," Ronald Reagan reminded us every Sunday evening on behalf of GE, "is our most important product."

It happens that as I write this some twenty-three years later, I am sitting in a stuffy conference room with three GE managers, two lawyers, and six other consultants, rehearsing a presentation that we will make next Monday to the Nuclear Regulatory Commission, which has shut down GE's reactor at Vallecitos, pending resolution of certain questions that have been raised about its ability to withstand earthquakes.

The meeting will probably go on all day. These days most forty-year-old technologists like myself spend a good deal of time in meetings like this. It is an appropriate setting in which to consider the professional relevance of what was learned in one's freshman year, two decades ago. First I note that as usual at these conferences on Vallecitos, the atmosphere is grim, even angry. The reactor, sole U.S. source of certain medical isotopes used in cancer treatment, has been shut down for three years now. For GE, it has been three maddening years of meetings, geological investigations, structural analyses, conference calls, legal fees, hearings, and consultant bills, trying to get the reactor restarted. A crack GE operating staff, built over two decades, has mostly drifted away or been laid off. Our country now buys its isotopes from a foreign government. The issues and the paperwork have metastasized. Careers and ideals are on the line. Our children are eyeing us suspiciously. "Harnessing the forces of nature" has become a kind of war crime. "The policy of this country is being taken over by women," one of the GE engineers says to me, almost in tears himself after a hearing in which a young mother has accused him of poisoning the water supply. We feel wounded outrage. "They have perverted their technical integrity," someone remarks of members of the U.S. Geological Survey, consultants to the NRC, who are holding press conferences and openly conferring with the lawyer representing the antinuclear Friends of the Earth. The feelings on both sides run deep. Once after a hearing, I ask the Friends of the Earth man whether there is any level of risk he is willing to accept. He says, "We don't accept that numbers baloney. You can't put value on the life of a child. GE's an ugly beast." He asks me who I am, and I tell him I'm a consultant to GE. "That's too bad," he says. Harsh words. I wonder whether there is not more lurking in the basement of everyone's mind than the nominal issue of seismic safety of a small test reactor.

One of the engineers, who holds a Ph.D. in structural engineering, is discussing the theoretical effects of an earthquake on the comatose reactor. The USGS claims that the reactor is sitting on an active earthquake fault. A "possible" fault. That there "could be a fault." Their claim is vague. The fault might move beneath the reactor, lifting it off the ground. Could that happen? Maybe the fault would just go around the reactor. That is what happened to that bank vault in Guatemala City, isn't it? Several people look at me. I tell them that we have not yet analyzed that; if there were a fault, we should consider whether it could intersect the right half of the foundation. Lift the right end of the reactor. Push the soil around somehow. Anyway, here we have the reactor spanning there, then we have the ground shaking, the reactor shaking, superimposed on the perched situation.

I remind myself that the big questions, the confusing questions, the issues that will be argued and questioned at the hearing by these distinguished experts, members of the Advisory Committee on Reactor Safety, retired deans and writers of standard textbooks, are the freshman questions, not the graduate school questions. What is moving what, and where is it moving to? How do you define and set up the problems that need to be analyzed? Will the reactor move with the ground? Or does it matter? What will the reactor and the ground do when the shaking begins? Is it worse if the shaking begins while the fault is still moving, or can the two problems be dealt with separately? What are the net forces on the base slab of the reactor? No doubt that highly detailed mathematical modeling of the whole problem will take us into the most advanced areas of soil mechanics, seismology, engineering mechanics, not to speak of structural engineering. But the key to the problem, the decisions as to what model to use in the first place, are freshman physics homework problems. Forces on a body. Inertia. Friction. So here it is, twenty-three years later, and the most important thing I know is freshman physics.

But what about this fault, the fault that is, or might be, or could be? The NRC geologists say that the reactor is "within an active fault zone," that it "should" be designed to withstand the effects of fault ground movement, that one meter of movement would be "appropriate." What kinds of statements are these? Is that "should" something more than an emotive appeal? That "zone" a priestly incantation aimed at mystifying the lay person? Are these not as much matters of definition, of linguistic analysis, as of engineering? Are these not the very problems we studied in "Introduction to Philosophical Analysis"?

Or consider this. Based on careful inspection of a trench alongside the reactor, the geologists and soil scientists have concluded that radiocarbon-dated soil strata beneath the reactor have not been disturbed by faulting for the past 100,000 years. What then are the odds of a fault occurring for the first time during the remaining life of the reactor? Knowing that a very large bag may contain some black balls (a fault could occur) but that the last 100,000 draws have produced only white balls, what is the probability that the next draw will produce a black ball?*

The purpose of this debate is to determine whether the reactor really needs to be designed to resist fault ground offset or whether the possibility of geological faulting beneath the structure is so remote—say one in a million a year, or less—that the hazard may reasonably be dismissed as insignificant. Nothing is risk free, our rational argument goes, and it is hardly rational to trouble ourselves over conditions that are two orders of magnitude less likely than other known unavoidable risks (the risk of core meltdown, for example, is said to be as high as one in ten thousand per year). Hence we engineers are inclined to

---

*Following our presentation to NRC, several independent agencies were commissioned to review this question over a period of months. Results were presented in several voluminous reports. The answer, everyone agreed, was 1/100,000, more or less.

dismiss the concern over earthquake faults as being irrational or uninformed.

We have at this point descended through two levels of the problem. The first, the engineering aspect, defining what would happen to the plant, if the fault actually existed, was a matter of engineering analysis, the core of which is a homework assignment in freshman physics. At the next level, we considered whether it is reasonable to consider the hazard in the first place. This seems to be a matter of common sense, philosophical analysis, and elementary probability theory, all leading to some quantitative assessment of risk, which can then be compared to some acceptance criterion.

And yet, to our disgust, we find that the argument is not solved! "Yes, yes," people say, "that's all very nice, finite element and Bayes theorem, your typical engineering approach. But."

But what?

But you are not giving the right answer! Because you do not understand the problem. You do not even understand what an earthquake is! Consider that this fictional earthquake is really an artifact of our collective imagination. Consider that your idea of an earthquake is narrow and specialized. It concerns only the *physical* earthquake, ignoring the *mythical* earthquake. Attend carefully to the commentary of citizens who listen patiently to your scientific and engineering explanations.

FIRST CITIZEN (an elderly woman): Mr. Chairman and honorable members of the hearing committee, I would like very much to speak to some matters that have been bothering me for a long time since the current attempts by certain Back to Nature advocates to push us backwards, in fact. Believe me, I was there, and many like me were there, too; backward, that is. The actual facts weren't pretty, healthy nor easy. Contrary to "The Little House on the Prairie" on television, those householders were not marijuana millionaires from Mendocino, the ones who harvest their zillion-dollar crops sometimes out of a clearing in the natural forest. . . . These are the ones, the marijuana millionaires, who also find PCP, heroin and cocaine to snow us gently. The poor souls who follow their league have not found nothing but a drug dream. Those growers have generated more deaths than have resulted from the nuclear fission process.

SECOND CITIZEN (a young man): The fact that last year we had three earthquakes on the Calaveras Fault and this year we've had a series of earthquakes in the Livermore Valley, indicates that there is a new trend for seismic activity on the Calaveras Fault. . . . Anyone with any sense knows that it's not safe to operate a reactor so close to an earthquake fault. And this reactor over here is less than a mile from the Calaveras Fault. . . . So I think that it's incumbent upon you to fulfill your responsibilities . . . to consider these recent earthquakes that we've had and the implications that it would have if this reactor was relicensed to start operating again and we have that 7.5 earthquake on the Calaveras Fault, and it tipped over that reactor or it split it open or just broke all of its pipes and released that radiation into the Bay area.*

What kind of dialogue is this? What do they, the people, mean by "earthquake"? What are these citizens talking about?

To contend with that question, I descend to a third level, to the basement of our culture, to that remaining freshman course, "Foundations of Western Civilization." Soon enough, I find several earthquakes there. In Homer, Thucydides, Euripides.

One of the most interesting of these quakes is an important event in Euripides' play, *The Bacchae*. It occurs when Dionysus, long-haired god of the counterculture, has been thrown into jail by Pentheus, the authoritarian king of

---

*Testimony recorded at the ACRS hearing a week after our meeting.

Thebes. Timothy Leary jailed by J. Edgar Hoover. Dionysus has had his fill of what he considers Pentheus' police brutality. Suddenly he shouts from his jail cell, "o dreadful earthquake, shake the floor of the world!" And the chorus replies:

> Pentheus' palace is falling, crumbling in pieces!
> Dionysus stands in the palace; bow before him!
> We bow before him.—See how the roof and pillars
> Plunge to the ground!—Bromius is with us,
> He shouts from prison the shout of victory!

By the end of the play, Dionysus, hardly content with a seismically aided jail escape, vengefully arranges to have Pentheus brutally murdered by his own mother, a convert to the Dionysian cult.

Readers of *The Bacchae*, including learned academic critics, have traditionally sought to draw a moral from this, the last of Euripides' plays. Political conservatives such as the First Citizen have seen the play as a sober warning against the excesses of nature-worshipping cults and emotionalism. Romantics, such as the Second Citizen, see it as a condemnation of Pentheus' emotional repression and smug establishmentarianism. It is easy to find support for either view in various details of the drama, and that is perhaps why the work has intrigued the Western world for so long. Readers find in it a welcome echo of their own prejudice.

But what catches my attention is a striking resemblance of the ideological conflict of the Bacchae to the case of the seismic safety of the Vallecitos nuclear reactor. Vallecitos, nominally a matter of public safety, is in reality a collision of highly polarized ideologies; it is the magazines "High Times" versus "Fusion," the Greens versus the Whites, Rousseau versus Voltaire, the Aspen Institute versus the Denver Coal Club. Stripped of its contemporary trappings, the issue at Vallecitos is identical to the issue presented by

Euripides nearly two thousand years before the age of the atom or the corporation.

On the one hand we have the view that the Dionysian "marijuana millionaires" are luring us with false and sentimental dreams of going back to nature, a strategy bent on undermining our civilization. On the other side of the issue are the Dionysians: do not defy nature, they warn us. "Anyone with any sense knows that its not safe to operate a reactor so close to an earthquake fault," says the Second Citizen in 1980. "Quickly will Pentheus' halls be shaken apart in collapse" echoes the Greek chorus in 405 B.C.

Scratch the surface of the Vallecitos dispute, of the whole issue of nuclear safety, of much of the present-day environmental controversy, and you find a social and political issue as old as Western civilization, a quarrel that predates and underlies current debates over corporations or radiation or the ecology, that transcends our technology.

In this light the earthquake safety of the Vallecitos reactor is no more the real problem than the seismic safety of Pentheus's palace is the real subject of the Bacchae. In both cases the real issue is ideological conflict, little changed in two millennia. Symptomatic, perhaps, of the disintegration of a golden age of democracy. It is perhaps little wonder that we have difficulty in understanding each other, we engineers and you Friends of the Earth, for we are talking in metaphors, defined for us less by ourselves than by the events of our time. Our earthquake is a powerful but stupid beast that can be tamed with the sweet songs of Newton and Bayes. Your earthquake is a different matter; it is the almighty rage of a god.

All men are tempted. There is no man that lives that can't be broken down, provided it is the right temptation, put in the right spot.

Henry Ward Beecher,
Proverbs from a Plymouth Pulpit (1887)

By all accounts I have heard, Thomas Worcester was a gentleman and a fine engineer, a worthy successor to his father, who some years ago had established the Boston consulting engineering firm of Thomas A. Worcester, Inc. Indeed, most New Englanders are well acquainted with the defunct firm's work without necessarily being aware of it, for Worcester's green steel bridges and turnpike cuts and fills are a functional if inconspicuous part of the Massachusetts landscape. Worcester himself was a Harvard man, gray suited and dignified, hardly distinguishable from the investment bankers and lawyers who walk with brisk dignity on Federal and Milk streets. He was a proper Bostonian, what my family would describe, not without some admiration, as a Yankee. And a successful one at that, for after the war, when that particular breed had weakened and many old Boston businesses were on the decline, the Worcester engineering firm was booming, growing in the late 1940s from a few dozen to over two hundred employees.

The success of Worcester's firm during those lean years was based in part on the widely recognized quality of its work. Worcester's resume included $100 million of successful wartime government contracts, and one of his new Route 128 bridges had won a national prize for its design. The firm's success was less obviously but undeniably attributable to a man named Frank Norton, whom Worcester hired in the late 1940s. Norton's calling card described him as assistant to the president, and the assistance he provided concerned that most delicate aspect of the consulting engineering business, which those of us in it frequently describe to our clients as "business development" but discuss among ourselves as "getting jobs." As I well know, having been at it some ten years, it is a constant source of worry, for consulting work is subject to the worst feasts and famines and is highly competitive besides. How often we find ourselves having hired and trained a large and fine staff to complete projects in hand, only then to find the work dry up so we are faced with the prospect of laying off those people we have trained and to whom we have become attached. Or worse, watching the firm's debts pile up while we maintain our staff, anxiously hoping for the arrival of a big new contract. And of course, beyond the goal of corporate stability, there are those little prickles of ambition, the firm of one hundred employees instead of only ten, the invitations to be keynote speaker at conferences or to testify at important hearings, pricey schools like Milton and Beaver Country Day for the children, the teak sailboat at Marblehead.

According to well-established canons of professional behavior, civil engineering consultants, like architects and lawyers, are supposed to be selected for public work on the basis of quality and reputation, through a gentlemanly process of discussion and negotiation, not by grubby competitive bidding. After all, the cheapest fee is not likely to yield the best design. Yet it is sad but true that in the absence of bidding, the selection process sometimes boils down to a matter of personal influence, contacts, favors, and even worse.

So when in the late 1940s a highly political official named William F. Callahan recaptured the post of Massachusetts state director of public works, traditionally known as a most powerful position because of the broad authority of the director in awarding contracts, it became essential that firms like Worcester's establish contacts with Callahan. And Frank Norton was the link, for as Worcester would later recall in court, Norton was "certain he could get the jobs for me because he knew Callahan." Norton was hired at a good salary and given a generous expense account. There are, of course, certain expenses that go along with such a key position as assistant to the president— dinner with clients and cronies at Jimmy's Harborside, an occasional trip to Florida to work out details.

And just as he had promised, Frank Norton began to deliver the jobs in the form of contracts from the Massachusetts Department of Public Works. One was a real plum: the design of the first eighteen miles of Boston's new peripheral highway, Route 128, which included forty-four bridges.

Meanwhile Frank Norton's expenses were as impressive as the contracts he brought in. During one period when Norton delivered $2,750,000 in new engineering work, his expenses amounted to $275,000, or exactly 10 percent of the contract amount. True, sales costs of 10 percent are not entirely out of line in most businesses, including consulting engineering. But what was troublesome was that Norton needed the money in cash, vaguely promising but never providing an account of its disposition.

As owners of small firms well know, getting hundreds of thousands of dollars in cash out of a company is difficult unless the cash is recorded as going to some legitimate subcontractor, supplier, or landlord or unless it is paid out as taxable salary or dividends to the company officers. The company check registers are, after all, subject to audit by IRS tax agents, who are not likely to accept, without further explanation, some $275,000 in checks written to petty cash or to someone's cousin for interior decorating services. Like so many others who have found themselves in this awkward position, Thomas Worcester began to "borrow" a little money from the firm here and there, to tide things over. There was, of course, no going back, and it was not long before the distraught engineer was writing corporate checks to companies and employees who did not exist, then cashing them himself, delivering the bundles of tens and twenties to Norton. Did Worcester, we ask now, really know what was happening to this money? From what he later told me, I would assume he did; but Norton was good enough not to trouble Worcester, a gentleman, with unsavory details. He just delivered the jobs, as they had agreed.

If Thomas Worcester will emerge from the end of this story ennobled by tragedy, Norton is but a minor player, dead now twenty-seven years. But as I reflect on the fragments of court testimony that I have read and on the story that Worcester himself told me one rainy afternoon in his office, I am compelled to pause for a moment, for I feel an odd sympathy for Francis C. Norton, whose name is, I suppose, appearing here in print for the last time ever. In these righteous, post-Watergate days, we are much inclined to moralize, perhaps superficially, on the conduct of men like Frank Norton. But perhaps the Nortons of the world smile at our moral naiveté. Perhaps they say to themselves, "You, hypocrites all, are comfortable with your lying and thievery, because they conform to the standards of your peers. But we see you in ways you would never dare to see yourselves. For we are the boundary mediators, ambassadors, Indian scouts. We alone have the courage to slip through that perilous no-man's land between your petty clans. It is a lonely calling, but not without honor."

There is a detail of Norton's demise that touches me. There had been a minor stroke, a chronic heart condition. Frank was not in good health. The pressures were slowly bringing him down, affecting his arteries. A fragment of court testimony harshly illuminates the beginning of the end. The scene was a friend's house, a Judge Charles Flynn's, a few days after Christmas 1951. A children's Christmas party went on in the background, and Flynn poured Norton a stiff drink. Norton had hired Flynn as a lawyer to represent the Worcester firm. The IRS had begun to breathe down Worcester's neck.

"I feel better," said Norton, sipping the drink, his second.

"It's going badly," said Flynn.

"For who?" asked Norton.

"You know for who. It's going badly for the Worcester Company. And for you."

"For me?"

"Yes."

"Well, I'll take care of that," said Norton, rising to leave.

Who among us does not know how sad New Year's Day can be when matters go badly! I imagine Norton sitting at his dining room table, that first day of 1952, staring at his tax returns. It was going badly, and it was up to Frank to take care of it. And he did. Mrs. Norton found him there later that day, slumped over the papers, dead of a heart attack.

If Thomas Worcester had felt some discomfort at the act of passing money to Norton, he must have found it that much more distressing to wrap up the bills in an A&P grocery bag and pay a call to the Norton household. For with or without Frank, certain scheduled payments were due nonetheless; without Frank around and until some other arrangements could be made, it had been decided that Mrs. Norton would have to handle matters. As Worcester had expected, there was another guest at the Norton home when he arrived one evening on his distasteful errand, and although he never saw the man—he had stepped out of the room for a moment when Worcester arrived— Worcester testified that Mrs. Norton had explained that the

54   overcoat hanging in the hall, into which she put the packet
of money, belonged to William F. Callahan.

As time passed, the fat public contracts on which the
Worcester firm was now becoming dependent began to
bring other obligations, other favors for Worcester to perform. There were the out-of-office politicians like Ed
Rowe. Somebody in high Democratic party circles had decided that Ed Rowe should be a Republican candidate for
governor, a spoiler, to split the Republican election effort;
to further this, Worcester hired Rowe for $20,000 to assist
in representing the firm. And certain political lawyers,
like Henry Santuososso, who had to call Worcester's office
one day to ask what the firm's records showed his $10,000
legal consultation was supposed to have been about.
A certain city councilor, who was paid $50 a week for
seventy-seven weeks for doing nothing. A firm called
Public Relations, Inc., hired for $29,000 to "stand by."

One day, an associate would later recall, Worcester just
buried his face in his hands. "It's got to be stopped if we're
going to keep business going . . . it's almost impossible to
do business today without somebody looking for something," said the unhappy engineer.

If we're going to keep business going. The responsibility
for enterprise, that moral imperative accepted so eagerly
by those engaged in business, entirely overlooked by those
who are not, with the result that this fundamental and
recurrent rationale is almost always overlooked by both
sides in any debate of the ethics of shortcutting the legal
and bureaucratic system.

Those readers who are unaccustomed to the ways of big
cities in the East should recognize that what I have described so far was, and for all I know still is, the normal
manner of doing public business in Boston, just as it is and
probably always will be in Bangkok and Honolulu and
Accra. This was no secret. As recently as two years before
Norton's employment by Worcester, James Michael Curley

had taken office for the fourth time as Boston's mayor. On election day, Curley was under indictment for his acts as president of an organization known as the Engineers Group, Inc., a business promotion enterprise started by an ex-convict friend of Curley's to peddle war contracts to engineers. The mayor's election-winning slogan was "Curley gets things done"; but whatever the mayor elect got done for the first five months of his mayoral term had to be done from Danbury prison where he was simultaneously serving a jail term for mail fraud. This was not the first occasion Curley had done time while in public office; indeed the veteran politician had capitalized politically on an earlier jail sentence—he had taken a civil service exam to help a friend—and his authorized biography, *The Purple Shamrock*, which Curley gave as a gift to his admirers, indicated that "there wasn't a contract awarded that did not have a cut for Curley."

This kind of corruption was condemned by proper Yankee Boston, but the continued political success of Curleys and Callahans proved that the majority of voters romanticized their behavior, many considering the shakedowns of businesses like Worcester's as a kind of retroactive tax on past injustices perpetrated by the Boston brahmins on poor immigrants. Who couldn't remember a sweet Irish aunt or grandmother who had scrubbed the floors of fine houses on Beacon Hill? And wasn't that Yankee moralizing really just sheer hypocrisy? "They all paid him—the banks and the utilities and the rest," one of Curley's underlings would recall when Curley had passed from the scene. "But who were the ones who did the paying? All of them, *old Yankees*." Not only was Yankee justice, encoded in those abstract and unintelligible leather-bound tomes in which the evil spirit of Cromwell still lurked, basically slanted to favor the established rule over the people, but worse yet, even the Yankees themselves cheated when it suited their purposes.

If James Michael Curley remains in public memory as the last of the flamboyant Boston bosses, observers of the Massachusetts scene agreed that the state's real power broker during the postwar years was its elusively quiet but iron-willed director of public works, later chairman of the State Turnpike Authority, William F. Callahan. Callahan, into whose overcoat pocket Thomas Worcester said he had seen his money disappear, was Boston's Robert Moses, the man who really got things done, whose political power was such that he never lost a vote in the state legislature. The son of a shoe factory worker, Callahan had earlier lost his public works job during a shakeup initiated by the Republican Governor Saltonstall in the 1930s but recovered his position when the Democrats came back to power in the late 1940s.

Like most other men in our suburban Newton neighborhood, Bill Callahan had worked hard. He had built an organization that had accomplished visible works. "There is a certain pride in building bridges, building roads," he said. The Callahans lived in a fine white house; one of my secret schoolboy shortcuts went through a grove of forsythia at the back of their ample lot. He was a balding, jowly Irishman. A cancer operation left him without a larynx, but with great difficulty and characteristic determination he learned to speak through a hole in his chest. "In 1952 I had my throat cut by a doctor. I've had it cut in other ways many times since," he said. The Callahans had lost a son, Bill Jr., in World War II. He was inclined to spoil his daughter, Jane. She married a Johnny Kelley. Callahan hired a firm called Highway Traffic Engineers, Inc. to work on the Massachusetts Turnpike on the condition that half the firm's stock go to Johnny and that he be paid a salary of $20,000 a year. Kelley's principal contribution to the firm seems to have been attendance at the annual shareholder's meeting. The Kelleys lived in Wellesley, which is even nicer than Newton.

The several reformers who had set out to topple Callahan from his position discovered a crafty and determined opponent. One crime commission lawyer recalled the Callahan style: "He was able to see to it that business was placed in the hands of persons from all walks of life, sometimes on a very thin basis as far as return services were concerned, and he was able to combine the natural loyalty which arises out of a financial relationship of that kind with a particular skill in appealing to the variety of the persons who were the subject of his benefaction. His sources of communication were remarkable. He was able to find out what every man cared about, and somehow, directly, or indirectly, to appeal to those desires."

Snooping investigators were never physically threatened but they were audited, received strange calls and callers, fed anonymous leads and tips that later proved false, which led to blind alleys, and which made them appear foolish to the public or within their profession. "The shots that were fired by this remarkable man were of a subtle and perceptive kind and created an atmosphere of embarassment, confusion, and uncertainty which tended to sap one's resolve," said one prosecutor.

And yet William Callahan was said to live a clean personal life. He did not drink or smoke or even play golf. He was a man who loved work, politics, and family. His interest ran to power, and he made a clean distinction between power and knowledge. Of those critics, especially academic critics, who opposed his high-handed methods, he said, "You've got to have critics. But usually they're people who haven't accomplished anything in their lives. I call them grocery-store philosophers, pen pushers."

One of the critics Callahan most definitely had in mind was a lean, lawyerish-looking young MIT professor of civil engineering, A. Shleffer Lang, who in my junior year there had allied himself with some Harvard Law School people to oppose one of Callahan's pet projects, the extension of the Massachusetts Turnpike into central Boston. One can imagine and even sympathize with the aesthetic objection to ramming a huge, reinforced-concrete, traffic-bearing structure right into the heart of Richardson's and Olmsted's Boston. There is, it has always seemed to me, a fundamental issue of quality, of taste, to be considered in these matters. But it is a characteristic of our times that these public questions cannot be debated in terms of quality; perhaps such terms carry excessively aristocratic associations. It becomes necessary instead to demonstrate that the offensively crude is fiscally irresponsible or that it constitutes a health hazard. So in opposing the project, the professors used the same arguments of financial irresponsibility used by the Saltonstall commission that had fired Callahan back in the 1930s. In the spring of my junior year, their anti-turnpike campaign, which had culminated in a series of public advertisements opposing the project, successfully frightened Wall Street investment bankers from purchasing $175 million in bonds for the proposed extension.

If Callahan was furious with the interference of these "pen-pushing" Harvard and MIT professors, by the end of that same year he had encountered an even more determined enemy. This was Judge Charles E. Wyzanski of the U.S. District Court of Massachusetts, who had taken on the 1960 tax fraud case of *United States* v. *Thomas Worcester* after a previous judge, William McCarthy, had been too busy to try the case in the last three years before his retirement. Wyzanski's starchy disapproval of Massachusetts politicians and their methods was well known. The judge was fond of comparing, for didactic purposes in his prep school commencement speeches, the righteousness of Massachusetts's earliest governor, John Winthrop, with the depravity of its recent governor, Curley. Presented with a tax-evasion conviction of consulting engineer Thomas Worcester, Wyzanski seized this opportunity to "sound a clarion," as he described his own actions, to expose the corruption that was rotting what he imagined had once been a commonwealth of honorable men. He accordingly offered Worcester a deal: in exchange for suspension of the eighteen-month jail sentence that the engineer had received for evading taxes on $180,000 of illegitimate payoffs, Worcester agreed to testify with complete candor on the disposition of the "Worcester bounty," as the judge called the payoffs and favors that the engineering firm had so liberally dispersed in return for public contracts. Wyzanski's purpose was nothing less than exposing a statewide "network of corruption," which included public officials at all levels of the state, along with the other institutions—banks, insurance firms, the press, and the bar—extending even to members of the judiciary, for he believed that all of them were, with greater or lesser degrees of enthusiasm, participants in the system of bribery and extortion. Worcester, whom Wyzanski said had "completely persuaded me of his decency," agreed to testify in order to avoid jail, and for several months in the fall of 1960, Wyzanski, in an unusual series of hearings, used Worcester's testimony as a starting point for a major exposé of the corruption rooted in the Massachusetts system of public works, and most specifically in that perennial soft spot, the award of engineering and insurance service contracts by Callahan's Turnpike Authority. For three months, Wyzanski subpoenaed and exposed state officials, including Callahan himself, to the questioning of both Worcester's attorneys and the U.S. attorney, Elliot Richardson.

Predictably, Wyzanski's efforts were harshly criticized in some sectors of the press, perhaps even more so in the streets and bars, after Sunday mass and over corned beef dinners in Boston's predominantly Irish neighborhoods. The ethnicity of the name was unclear—was he a Catholic or a Jew?—but the affiliations were soon enough revealed. Wyzanski had attended the prestigious Exeter Academy, then gone on to Harvard. He was a member of Boston's

Signet and Tavern clubs, was prone to sprinkle his college and prep school commencement speeches with references to Pericles and Locke and Wordsworth, to reminisce about his Exeter friends, boys with names like Francis Plimpton and Dudley Orr and John Cowles. Whatever he was, Catholic or Jew, one thing was clear enough about this Wyzanski: he was trying to pass himself off as a Yankee.

Indeed they were all Harvard men: Worcester, Wyzanski, and Richardson. Elliott Richardson had been a legislative assistant to the same Governor Saltonstall who in 1939 had dismissed Callahan from his public works position for squandering public funds. Richardson's career and affiliations had been a carbon copy of Wyzanski's; both had graduated from the same school, clerked for the same justice, Learned Hand, and been associated for many years with the oldest Yankee law firm in the city. Were these not grounds to think that Wyzanski was Richardson's mentor?

Surely this highly publicized hearing, initiated by Richardson and blessed by Wyzanski, had a larger purpose than adjudication of the trivial question of whether Thomas Worcester's probation should be revoked. Clearly it was an attempt to try the political system of the state, an assault on the institutions of family and church, a campaign to undercut the personal loyalty that protected people from exploitation by the traditionally powerful Yankee establishment. Wasn't this only an extension of the communist subversion of American ideals, which—as Senator McCarthy and others had shown in the early 1950s—was infecting institutions like Harvard and MIT? And who could ignore the fact that it was an election year, that Bobby Kennedy was already searching Cambridge for candidates for high-level federal posts? Wasn't Wyzanski, who had served with John Kennedy as a Harvard overseer, really just creating a little election year publicity for himself, perhaps bucking for a Kennedy appointment? Could

it not be said that behind that Harvard mask of righteousness there lurked more in the way of personal ambitions
than Judge Wyzanski would care to admit?

If men of action like Bill Callahan cut corners to get
things moving, took care of their families and friends,
spread the work around to the small people, what of it?
Consider that Callahan had been accused, in which his
public relations man called "Mr. Wyzanski's Nazi trial," of
pushing turnpike insurance business toward friends
among the fifty-four insurance brokers who were members
of the state legislature. But why shouldn't he? Callahan
pointed out how when the Republicans were in office, the
big, old-line insurance companies got the business. "The
little fellows didn't get a look. Somebody has to get the
insurance business. And it don't cost you any more
whether you insure through John Smith or Mike McCarty."
The difference was that Mike McCarty, barely getting along
with six kids, would remember the favor when it came
time to vote. John Smith—or Thomas Worcester—offered
money, not loyalty.

"Judge Wyzanski has long enjoyed a reputation as a brilliant jurist; having been before him a few times I can attest
that he has a piercing tongue and steel trap mind," writes
a Boston lawyer friend of mine, enclosing a copy of the
judge's opinion on the Worcester proceedings, delivered
to the court on December 29, 1960. The twenty-eight-page
document is a work of dazzling erudition, which draws
upon the imagery of Melville, the Bible, the Koran, the
wisdom of Heraclitus, Voltaire, Thoreau, Johnson, Plato,
Whitehead, G. E. Moore, Montesquieu, Sophocles, Shake-
speare, both Huxleys, James, Emerson, Webster, and San-
tayana, not to mention dozens of legal authorities.
Wyzanski, mindful of the accusation that he is pushing his
name, his views, his personality before the public in an
election year, concludes the opinion with a characteristic
flair:

But even if one is to be charged with vanity, with absence of taste, with lack of grace, with lust for higher office, is it not time to sound a clarion? Another will blow it better. But it is worth something to prove that the trumpet can be blown.

"To do justly, to love mercy, and to walk humbly with thy God" is a command which sometimes can be fulfilled only in a spirit of righteous indignation.

The hearings are terminated.

That same fall I had finally reached the relatively safe anchorage of my senior year at MIT. It was a presidential election year, and for those of us from Boston a year of paradoxical choices. Presidential candidate Richard Nixon, coming from that land of pinko lotus-eaters, California, was a staunch anticommunist. John Kennedy, grandson of a Boston mayor said to have been even more corrupt than Curley, was a liberal who sat with Judge Wyzanski on Harvard's Board of Overseers. And yet it was Bill Callahan, not Charles Wyzanski, who was invited to Kennedy's private birthday party; Callahan sold most of the fund-raising advertisements on the souvenir menu. Neither presidential candidate measured up to my standards of honor. Friday evenings, at my proletarian hangout, the Paradise Café, I defiantly played the pinball machine instead of watching the televised debate. Why vote, I would say to my parents, when the choice is between two phonies like that?

It was the year in which I was too cynical to participate in the new dance craze, the twist, but earnest enough to criticize our unfair play toward Castro's new Cuba. Like most of my classmates, I was resigned to compulsory military service after graduation, but there would be plenty of good jobs waiting after that. It was 1960, and *progress* was still an important word. The first commercial nuclear power plant had just been licensed. A couple of young engineers were starting a new company called Teledyne.

The laser, the felt-tip pen, and Enovid were perfected that year, and the space race would go on. Few things were as bright as the future of technology.

With the exception of one structural design course, MIT seemed easy to me that fall. I wrote a FORTRAN computer program that aimed to beat Fats, the bookmaker at the Paradise (it did not); studied the strength of the building materials of the future, foam core aluminum panels and fiberglass-reinforced plastic; broke asphalt cylinders in the basement of Building One. For the fun of it, I took two humanities courses. One was taught by a wickedly funny visitor, John Hawkes, whose novels were just coming out as New Directions paperbacks. The other, a course in moral philosophy, was taught by Yale philosopher John Rawls. I did not understand much of the scholarly Rawls, about whom a book has recently been published, *Understanding Rawls*. Rawls's spare, bespectacled appearance and scholarly manner were just what one would expect of that endangered species, the moral philosopher. Most of the specimens, one gathered from our assigned readings, were to be found in English academic preserves, tut-tutting each other's "fallacies" at croquet or over tea. I soon read enough of G. E. Moore and A. J. Ayer to realize that they would never presume to replace the authority of the church fathers, whom I had chosen to dismiss a few years earlier, thereby creating a gap in my moral sensibility as troublesome as a missing tooth. It seemed to me that the principal concerns of modern ethical philosophers were linguistic; they wanted to define precisely the meaning of the word *good* in "Bill Callahan is a good man" or *should* in "Thomas Worcester should have demanded an account of Norton's expenses." I gathered that one school of philosophical thought had gone so far as to declare that such moral statements had no meaning whatsoever; *should* and *good* in their view were no more than animal grunts of approval. If this were true, then of course these Harvard lawyers and Boston Irish pols were no more than two clans of peeled apes, screeching at each other across the Charles River, and Thomas Worcester's only wrong was the practical mistake of trying to operate across tribal boundaries. Or perhaps, in accordance with Immanual Kant's basic and to me more sensible rule, a man would be right provided only that he acted as if he willed his actions to be universal laws. In that case, wouldn't Callahan and Wyzanski, each of whom would be content presumably to live in a world of their own values, both be logically "right"? Then only Worcester, whose respectability did not match his conduct, would be wrong. Could it be that the existence of diverse hardy ethnic subgroups, each with its own "operational code," provided a benevolent ecological stability to Massachusetts, a political state which might be preferable to mass anomie, to spells of weak leadership broken by periodic revolutionary violence?* Was honesty to be expected in a true leader? "Our rulers will find a considerable dose of falsehood and deceit necessary for the good of their subjects," said Socrates. It was all very confusing, from an ethical standpoint. Now in leafing through my twenty-year-old text for Rawls's moral philosophy course, Sellars and Hospers's *Readings in Ethical Theory*, seven hundred pages sturdily bound, I am pleased to find that I underlined the following statement, which still best expresses my retrospective findings on the practical applications of moral philosophy: "It is therefore very easy when we investigate these matters philosophically to go wrong because we do not have before us at the time a genuine ethical experience, and at the very moment when we have such an experience we are too

---

*Readers interested in a sociological view of public bribery might enjoy W. Michael Reisman's book, *Folded Lies; Bribery, Crusades, and Reforms.*

65   much concerned with it as a practical issue to philosophize
      about it."

Spring term of 1961 was my last at MIT, and as the putty-
colored limestone buildings of the institute began to warm
in the April sun and the grass of the quadrangle grew thick
and green, I began to feel a sense of lightheartedness that
I had not known since high school summer vacations. For
me the struggle for grade-point averages was over; the last
lion, "Structural Analysis and Design," was replaced by a
lamb, "Social and Political Factors in Engineering." Said
to be a gut course, it was taught by the same Professor Lang
who earlier had tangled with William Callahan on the
turnpike extension bond issue. Lang's aim was to expose
us to some of the nontechnical problems that engineers
are famous for not recognizing. I had been following
the Worcester case in the newspapers, and when I suggested that as a course project I undertake a study of the
scandal, Lang told me that would be an excellent idea.
I accordingly called Worcester's attorney, Calvin Bartlett, and explained my interest in the case. Bartlett invited me to his office and gave me a large cardboard box
full of court transcripts. When I asked him a question
about Worcester's attitude, Bartlett suggested that I ask
Worcester himself, picked up the phone, and called the
engineer. Would Worcester talk to me? The answer was
apparently yes. Bartlett asked me what day would be
convenient for me the following week and set up an
appointment.

So it happened one rainy spring afternoon that I took
the MTA downtown to visit Thomas Worcester. I was nervous and wore what seemed to me to be an appropriately
funereal charcoal suit and dark blue tie. It was a blustery
day and by the time I found the correct street address, my
London Fog raincoat and suit pants were soaked.

In the cramped lobby of one of the older buildings in

Boston's financial district, I found one of those glass-en-
cased, black felt building directories on which press-on
letters indicated that what remained of the Worcester
firm occupied the fifth floor. A fifty-foot ascent in a creaking elevator and a short walk down a linoleum corridor
brought me to a heavy oak and glass door. Inside a male
secretary greeted me suspiciously, then escorted me
through a series of empty offices with dusty drafting tables
to Worcester's office. Worcester greeted me like a gentleman, and we sat talking for an hour or so. My recollection
is of a tall, stooped man with a starched white shirt, who
seemed tired but glad to tell me the story, much as I have
told it here. It seemed to me then as it does now that
Worcester, of all of them, on a purely logical basis had
committed the most grievous wrong. And yet Worcester
that afternoon had a certain stoic way about him, a certain
dignity that elevated him, the one man who had actually
been convicted, above all of the others, including the righteous Wyzanski and the determined Callahan. It was, I
suppose as I look back on it now, residual traces of Catholi-
cism in me that made Worcester's judgments carry a certain high authority, for only he among the many sinners
had done penance. I believed, moreover, that Worcester's
trial had yielded a good, that Massachusetts politics
would never be the same.

Only once that afternoon did that cool Yankee seem bitter, and that anger was directed at his fellow engineers,
who had put him out of the Boston Society of Civil Engineers for his conduct. "They denounced me and yet they
all conduct their business the same way," he said.

After we talked, Worcester accompanied me to the door,
we shook hands, and I walked out into the thickening
gloom. Worcester and the male secretary had been the only
people on the fifth floor of the building. Walking back to
Park Street, past the slate stones of Granary Burying

67   Ground glistening black in the rain, the realization came
       to me that Worcester's only reason for coming to the office
       that day may have been to talk to me.

Leadership traits are personal qualities which, if shown in
your daily activities, help you earn your men's respect,
confidence, willing obedience, and loyal cooperation. A
study of the lives and careers of successful commanders
reveals that many of the following personal traits or qualities are common to all of them.

Military Leadership
Department of the Army Field Manual FM 22–100
December 1958

1960   Integrity

There is no possible compromise.

Seventy of us, ROTC officer candidates now in our second
week of summer camp, sat on bleacher seats in the Virginia
sun. Sweat trickled beneath our T-shirts, and a kind of
diaper rash bloomed on the backs of our legs. My brain
roasted in the moderate oven of a helmet for three hours
while the sergeant taught us sight picture and trigger
squeeze. It was an accelerated course specially developed
for ROTC engineers; ordinarily a full day was devoted to
just the sight picture. For my part, three hours sufficed, for
twenty years has not worn away the image baked into my
nervous system. I close my eyes and there it is: the black
circle framing the front sight of the M1, the bull's-eye resting lightly on its square top.

   That evening before taps, the barracks clattered with the
sound of the Garand bolts, and we curled over our GI
boots, rubbing them slick with spit until the fat toes
glinted like eyes and the black Kiwi polish stained the tips

of our index fingers. Saturday morning was inspection time, and if everything went right, tomorrow they would give us our first weekend pass.

The next morning after breakfast we cleaned the urinals and searched the gravel paths for butts and rolled our brown army socks into funny-looking balls with lips. Mulrooney was assigned to the bunk next to mine. Physically Edward Ignatius Mulrooney consisted of a four-and-one-half pound liver plus 250 pounds of standard equipment packed into a pinkish freckled casing the size of a double bed sheet. We used to drink beer together evenings at the North Post Annex of the Fort Belvoir officers' club. Mulrooney came from one of those Boston Irish families that believed "to drink" meant "to get drunk." We often shared hangovers. Under those and other times of adversity Eddie always took on the responsibility for cheering us both up with jokes, whimsical fantasies invented during our calisthenic daily dozen, or sarcasms muttered while we stood in formation, and I was grateful to him for that. But Mulrooney was incapable of arranging his socks and toilet gear the right way in the upper tray of his footlocker, so I did that for him. The previous Saturday Eddie, while admiring my work, had picked up his can of aerosol shaving cream, put it back standing upright in the tray, then sat on the lid of the locker, causing the tray to fill with Gillette Foamy. This condition was not discovered until inspection, when Lieutenant Schimpf, our company exec, opened the locker in his quest for sock misalignment. So this Saturday I told Eddie not to go within three feet of his footlocker after my final arrangement was complete. "I don't want to think about you hanging yourself here while the rest of us are getting gooned in Alexandria," I said.

At nine-thirty in the morning the third platoon's barracks next door fell silent, which meant that the inspecting party had arrived there. We were next. We checked the blousing on our fatigues, trimmed thread ends from buttons and collars with fingernail clippers, stretched our brown blankets drum tight. Bedigian, our forward observer, stuck his head through the door. "Here they come, balls and all," he said.

I was standing next to my footlocker, looking at the wax-swirled wooden floor, when I noticed the thread. It was at least two inches long, a fat hairy one like the ones you found on the belt loops of new khaki pants. Footsteps were crunching on the gravel outside, and there was a last minute scramble followed by silence on the first floor, then the TEN-HUT announcing the arrival of the inspection party. Picking up the thread, I looked wildly for a place to dispose of it. Under the bed? In my pocket, starched perfectly flat? No. In my footlocker? Too late to hide it there. The butt can? Yes! THE BUTT CAN, that number 10 fruit can painted fire engine red, shined inside and out, filled exactly half full with fresh water, hanging on a post between Mulrooney's and my bunks. No one was looking. I dropped the thread into the water and stood at parade rest next to my footlocker.

Captain Edwards, our commanding officer, was easygoing; he asked semi-intellectual questions like, "Do you know why we have you officer candidates spend a full day digging a ditch?" (Answer: So you'll know how much digging a man can do in a day.) But his exec, Lieutenant Schimpf, with his bony shaved skull and glassy eyes, was a real screamer. That was one of those elementary principles that they did not teach in military leadership courses but that I learned by observation and never forgot. A commander can afford to be lovable provided he has at least one snarling bastard like Schimpf around to keep everyone in line.

Schimpf was leading the inspection party, and when he strutted up to Mulrooney and wheeled to confront him, naturally he had to bring up the shaving cream incident and make poor Mulrooney repeat his general orders "so someone could hear them," and so forth. And naturally, Schimpf decided, for what I suppose was the first time that day, to peer into that butt can, and what he saw made him freeze in his tracks.

"What is this, now, Mulrooney, you store your old clothing in this receptacle? Come over here; you put these combustible materials in this device that's intended as a fire prevention measure?" Schimpf now had the thread hanging from his finger, dripping onto the floor.

"No sir," said Eddie, loud and clear.

"You don't *want* to go on pass this weekend, do you?"

"No sir. Yes sir."

"Well if you didn't do it, who in this squad made this filthy mess that's going to cancel the whole platoon's pass unless he speaks up?" Schimpf's reptilean gaze fixed on me for a long time, then swept down the line. Confession rose in my chest like a bubble, then something twitched shut. I did not want to spend the weekend alone in that forlorn building.

Silence.

"No point in even continuing this inspection," Schimpf said, strutting from the room.

Black disappointment flooded the barracks. Eddie sat on his footlocker and held his head in his hands, like an Italian. "Now who the fuck did that," he said to no one in particular. I felt sick to my stomach.

If we had had a little more than two weeks of experience in the army, we would have known that word would come down from company headquarters half an hour later that the fourth platoon would get a second chance. By noon, following a desultory inspection by the first sergeant, we had our passes.

As for that black cloud of disappointment, it was condensed, distilled, and presented in a small flask to me by whatever gods attend to the moral sanitation of the universe. Once over beers, Mulrooney alluded to the incident in a way that would have allowed me to confess. Eddie had a good heart and would have offered absolution by making it into a great joke. But I foolishly passed up the offer. And so I keep that flask still in a lower vault of my mind, in a cotton-lined drawer labeled, "There is no compromise."

1961 Courage

Study and understand your emotion of fear.
Back in the old days, before it was discovered that *we* were our enemies, it was widely held that *they* were our enemies. They, of course, were the Russians and that pack of East German Dobermans they kept kenneled in their front yard. They were going to bury us, Khruschev said, and when you were doing your time in the army, it was easy enough to imagine what he had in mind. One day you would be tooling down the autobahn in your jeep and the next minute you would be batting uselessly at a swarm of East German tracers, then lying in the ditch, like a full-page photograph from Life's *Picture History of World War II*, the numb wooziness spreading over your chest, calling to your mama, a sweet saltiness in your mouth. Or absurdly trying to remember, after your first whiff of nerve gas, just before you started flopping around like a fish in the bottom of a rowboat, the correct procedure for requisitioning atropine from the battalion supply sergeant. Or perhaps we'd just be walking with the baby across the post exchange parking lot, five miles from ground zero, when there would be a soundless flash, followed by the hiss and crackle of universal incineration, a last glimpse of each other's surprised and blackening faces.

But they let us know that our army was not just sitting around nervously wringing its hands. Our scientists were coming up with some ingeniously nasty surprises for the Russians too. One muggy Friday afternoon in August fifty of us, second lieutenant trainees in the army engineers' officers' basic course, learned more about one of them than we cared to know.

It was a special classroom, situated in a part of the post that I had not seen before. The windowless building was set in a dark grove of trees and surrounded by a chain-link fence topped by barbed wire. Guards with unusual uniforms checked our identification against a typed list of names and security clearances. Inside the entrance to the building another guard sat at a desk and checked us again before opening an inner door with a buzzer. This opened to a corridor, down which we were escorted by yet another guard, through a kind of hatch, and into a small amphitheater. We filed into rows of wooden seats—no talking please, gentlemen, no notes permitted. They counted us again, then closed the hatch-like door, sealing it with a large lever.

Major Henkel stood erect and frowning on the stage. "Good afternoon, gentlemen." "Good afternoon, Sir." "Gentlemen, this afternoon we are going to review the principles of deployment and operation of a special demolition device." The major stated its formal name, the usual backward army description, followed by the catalog number, which I always figured would be used to order a few more, after you ran out, or to replace the one you had got wet that would not work. Major Henkel was known to us as a high-level instructor, a specialist in demolitions and mine warfare. He had a crisp, authoritative self-confidence; I could more easily imagine him commanding a tank battalion than wet-nursing second lieutenants. He was a soldier, not a professor. I could tell that Henkel was teaching from the manual, not from any knowledge of the fundamentals; if I were to ask him whether the amplitude of a blast pressure wave decayed as $1/R^2$ at the ground surface, he would not understand what I was talking about and I would upset the timing of his presentation. But under the circumstances none of us was about to play that kind of game, and Henkel smoothly presented his material, clicking little magnetic signboards—"tactics," "authority," "requisition," his key topics—onto the blackboard as he lectured.

Had the device been available at the end of World War II, it would have done away with the Ramagen Bridge and everything for miles around it in a thousandth of a second, Henkel said. Fully armed, it weighed one hundred sixty-two kilograms and could be carried in the back of a standard three-quarter ton truck. At this moment, two sergeants appeared at the side of the stage, rolling a dolly covered by a black cloth. When the dolly was centered on the stage, they removed the black cloth, revealing a stainless-steel cylinder the size of an inverted wastebasket. Then at a nod from Major Henkel, the NCOs unscrewed some expensive looking wing nuts and lifted the cylinder from its base, on which was mounted a roundish device that looked like an automatic popcorn popper with a dozen spark plugs sticking out of it. The top half of this opened on a hinge, revealing a perfectly round, hollow interior. Major Henkel paused to let this sight sink in.

Fifty second lieutenants sat in absolute silence. Most of us remembered that day sixteen years before when President Truman had announced that the force of the sun had been loosed on the Japanese. I remembered my grandfather telling me that it turned the desert floor to colored glass. I caught the eye of Eddie Mulrooney, sitting next to me. "Jesus," he said softly.

Major Henkel now began to tell us how we would put this thing into the back of a three-quarter ton truck, then drive it to whatever place we wanted to vaporize. ("Just park it within a few blocks; that'll usually do fine.") I imagined myself parking the truck, shakily putting a deutsche mark in the parking meter, whistling, "Don't Sit Under the Apple Tree with Anyone Else But Me" to keep myself calm.

As it stood now the thing was harmless. But when you parked it where the division commander wanted it, then you had to arm it. Major Henkel looked at the senior NCO and said, as if on the spur of the moment, "Sergeant, why don't you bring the arming device out here for these gentlemen to see." The NCOs exchanged glances as if in hesitation, then the senior one said, "Yes, sir," and went backstage. A minute or two later he returned. He wore rubber gloves and carried a perfectly round, polished metal sphere the size of a cantaloupe.

Meanwhile Major Henkel was warming up to his topic, telling us how the enemy would never know what hit them, that it was a tactical device that could be authorized at division level by the man who really knew what was going on. "Just put the arming device in there, show them how it fits," he told the sergeant. The NCO carefully lowered the heavy ball into position. "That will be all," said the major, and the two NCOs left the stage. Then Henkel lowered the top of the popcorn popper over the ball and fastened it tightly with what looked like four ski boot buckles. I could have sworn that he was humming softly to himself. He stood back, looking at the ugly device with a sort of paternal warmth. "Incredible piece of work," he said. I was beginning to wonder about Henkel. He was acting as if he had invented this diabolical engine himself.

The room, sealed up like a gas chamber, was getting stuffy. Some of us had been out at the officer's club drinking beer the night before, and I became aware of a cramp in my stomach. My mouth, seared by too many Pall Malls, tasted metallic, and my skin was damp and prickly.

Henkel continued. When the device was fully and cor-
rectly assembled and armed, we pressed this button—he
pressed it—which activated an amber light, confirming
that the device was fully armed. Next to the amber light
was a panel on which were displayed three pairs of glowing zeros, representing hours, minutes, and seconds, re-
spectively. There were two methods of detonation. The
first was an automatic timer, which could be set by dial up
to 99 hours, 59 minutes, and 59 seconds. Henkel demonstrated by setting the timer to read 00 hours, 03 minutes,
00 seconds. Then he flipped another switch, the trigger, he
called it. You had to hold open a safety latch with one
hand and trip that switch with the other. A red light
glowed next to the amber light and the timer immediately
flashed 02 minutes, 59 seconds, 58 seconds, 57 seconds,
making a fussy little clicking sound. I remembered a story
that my uncle had told me about how on the Pacific Islands
they had to let the safety catch flip off their hand grenades,
then count "one hippopotamus, two hippopotamus, three
hippopotamus" before throwing them at the Japanese, because the Japanese were such fanatics they would just
catch them and throw them back if there was time left on
the fuse. I always wondered whether I would be one of
those dumb clunks that just stood there paralyzed until
the grenade went off in my hand.

The timer was now clicking 01 minutes, 45 seconds, 44
seconds. I could feel my heart rate pick up and my breathing getting shallower. I had to be missing something obvious here. Could it be that I had dozed off for a second,
when they were removing that ball? How could they let
this guy—he wasn't exactly a division commander—fool
around with a dangerous thing like this? Suppose the major was a little mentally unbalanced? I glanced around the
amphitheater at my classmates. Fifty second lieutenants
leaned forward in the seats, expressionless, intent.

There was one additional method of detonation, the major was explaining; that was by remote radio transmitter. In his hand he was holding what looked like a walkie-talkie. Like any other transmitter, it needed an antenna. Suddenly there was a metallic hiss and a loud clack; a three-foot-long spring-loaded, telescoped antenna shot out of the walkie-talkie. I could feel my shoulders move spastically an inch closer to my ears (the head, protect the head!) and I heard Mulrooney beside me utter a little sound, "Mom" it sounded like. Fifty diaphragms stiffened. All eyes were on the digital timer, which was now reading 00–00–45. The phrase, "Somebody should do something, somebody should do something," began to repeat itself idiotically in my head, to the cadence of the clicking sound.

Major Henkel was explaining about the transmitter, how it needed to have a little coded capsule inserted into the back of it. He did not have one here, of course—that had to come from the division commander himself; it was inserted in the back, under this little flap back here. The major was evidently having some difficulty with the flap. "This little release catch is *supposed* to be pushed down." The timer was down to eight seconds. "It seems to be stuck." Three seconds. Major Henkel bent over the transmitter, mumbling to himself. I distinctly remember, as clearly as I do my basic M1 sight picture, that fraction of a second before the explosion, when the device displayed a perfect row of zeros.

It was a sharp pop, like you'd get when you hit a whole roll of caps with a hammer, and from the top of that dreadful machine a small puff of smoke arose. It was a perfect little mushroom cloud.

"Oh my," said Major Henkel, "Did I forget to turn that off?" He looked down at his watch. "Well I see it's *exactly* five o'clock. Class dismissed."

1962   *Justice*

Search your mental attitudes to determine if you hold any prejudices and, if so, rid your mind of them.

I became an instructor and taught a class of enlisted men at the Army Engineer School. The idea was to teach them the fundamentals of soil testing for construction projects; at the end of the eight-week course, they were supposed to be qualified soil technicians. If the army ever had to build a road in Alaska or an airfield in Indonesia, one of my men would be there sniffing black lumps of soil for organic odor or adding water to red laterite gravel to determine its optimum water content. I had studied these matters at MIT and was much impressed with my own scientific expertise in them. Only a few years previously, building things from earth had been considered an art like cooking or gardening, or building Roman aqueducts or Gothic cathedrals, but now we were turning it into a science. Teaching the course was a valuable experience because I discovered that in fact I did not know how to build a road myself, although I could talk learnedly on the subject. I realized that I was in effect sermonizing on the art of love, with Kinsey as my text.

This was my first experience at teaching, and a first opportunity to impose my own social philosophy on others. I had attended a high school and university in which young men were educated along Darwinian lines. We had been given to understand that there were to be found in any worthwhile field of human endeavor two classes of people: the elite and the incompetent. A decade of this New Englandish philosophy had primed me for rebellion, and by the time I began to teach in the army, I had become infected with more romantic egalitarian ideas. It seemed to me then that all people were the same, from which it followed that the elite should be brought to their knees and the incompetent freed from their chains. I held to both this incorrect premise and its foolish corollary with the blind faith that characterizes all revolutionary thinking. In retrospect this inflexibility was good, for the obvious harm that came from this philosophy in the mid-1960s forced me eventually to recognize the danger of such social sentimentality.

It is a persistent irony that those who deny the existence of an elite and an incompetent are nonetheless fascinated by both and not the least interested in the ordinary. So it is that I am unable to remember any of the men in my class except Marine Gunnery Sergeant Roy Pickett and Private Henry Carver Williams. Pickett and Williams represented the two extremes, and the problems they raised provided me with an opportunity to apply for the first time the very principles that I believed should be enforced on a universal scale.

Sergeant Pickett's perfections far exceeded even the usual high standards of the U.S. Marine Corps. There were, of course, the usual Marine externals: knife-edged creases on gabardines, shoes a burnished lustre, the smoothly trimmed arc of cuticles. Pickett's manners and bearing were antiseptically faultless; he treated me, his superior in rank, with a polished respect that entirely concealed any feeling, if indeed he possessed any. Pickett's mind digested the technical material of the curriculum with apparent effortlessness; he understood everything the first time, and he had no questions or curiosity about why anything was done the way it was. No spiritual slouch, he was a minister in the Baptist church. I did not like Pickett, or perhaps I perceived and feared a core of sickness in him, just as a woman will fear a man whose dress and grooming are too perfect. In my heart I longed to make a little scratch in the polished surface that he turned toward the world.

Henry Carver Williams was, like Pickett, from Alabama.

Raised on a hard-scrabble clay farm in a house that lacked electricity and toilets, Williams had enlisted in the army to "improve himself." Williams may not have been too bright to begin with, but he had evidently received little schooling either. He read the way a third grader reads, and his writing was a childish scrawl. Nonetheless he attacked with admirable persistence the various testing and computational exercises that I assigned to the class, and it thrilled me to see him, as if in a painting by Millet, sitting before his little pan balance wrinkling his sweating brow and biting his lip as he attempted to memorize what to him must have been a score of unreleated mental steps necessary to compute the water content of a chunk of clay. It was the summer of Ross Barnett and James Meredith at Ole Miss. Like all other New England liberals of the day, I felt abolitionist embers glowing hot within and believed that racial hatreds would dissipate with proper instruction on such matters as the Boston Tea Party, speed reading, and the manual computation of square roots. Williams was a real find. Naive to the ways of romantic condescension, he was delighted with the special attention. The majority of his fellow students, all of them white, considered him a brown noser, which I attributed to their being rednecks, although I was a little troubled by the failure of a couple of them, midwestern college graduates, to endorse my promotion of Williams to the rank of noble savage.

I gave a final exam at the end of the course, constructed by cannabalizing multiple-choice questions from exams of years past, to which I added an original essay question: "A five-pound sample of soil has been delivered to your laboratory and you are asked to determine whether the soil is suitable as subgrade fill for a new road at Fort Belvoir. What tests would you perform on the sample, and why would you perform them?" The essay question was worth fifteen points. In accordance with the usual army practice of reviewing course work, I spent the last two classes giving the answers to all of the multiple-choice questions, but I did not explicitly cover the essay question, for I had tried hard to inculcate some judgment in my students, and I wanted to learn whether I had been successful in teaching them more than how to run tests.

As I had expected, the college-trained GIs did well in the multiple-choice questions and very well in the essay question, while the less-educated soldiers' performance was fair on the multiple choice and poor on the essay question. However, when I added up the final scores, all of my students but one had exceeded the minimum passing score of sixty, which was good because the army had a doctrine that said in effect that failure was the instructor's failure, not the student's. Only Williams, with a score of fifty-five, failed to pass.

I was disappointed. I believed that Williams had tried, really tried. Hadn't he told me that he had stayed up all night studying before the exam? "Ah'm jes gohn try mah best, Lieutenan," he had told me. He wanted to be a soil technician, unlike some of his classmates, for whom the army was just a two-year nuisance and who considered the whole business of soil testing (my chosen field) to be visibly primitive. Williams had the will to succeed. Wasn't this precious attribute, so sadly lacking in his classmates, worth a few points? Wouldn't the elevation of Williams from a hardscrabble sharecropper to a man who could faithfully act in accordance with the precepts of the American Society for Testing Materials yield goodness of such a quality to justify a minor adjustment of his test score? I riffled through the mental packet of index cards on which were recorded my conceptions and misconceptions of the moral precepts of great men. Wouldn't Kant agree that my humanistic flexibility was a desirable universal trait? Surely the Utilitarians would applaud the forthcoming social benefits of Williams's passing. And

Sartre would be the first to consider test scores a worse-than-useless metaphysical residue.

I reread Williams's answer to the essay question, what tests should be run and why. At first glance it was a meaningless catalog of whatever tests, most of them inapplicable and misdesignated, came to Williams's mind. Why run those tests? "Why is to test the soil," read his answer. I had given him zero of a possible fifteen. But perhaps I had been too hasty. After all, he had partially named two, that is one-third, of the six required tests. And why run those tests? Well, what could go straighter to the heart of the matter than "to test the soil"? Perhaps there was a sort of sly Oriental truth in this answer that was only too easily overlooked. We test the soil to test the soil. A rose is a rose is a rose. I erased the zero and wrote in five for the essay question, then erased the not easily erased red-penciled fifty-five and wrote down sixty. So it was that Private Williams became a certified soil technician in the U.S. Army Corps of Engineers and one of the earliest products of an affirmative action program.

That left me with that other awkward tail on the Gaussian curve, the one at the high end. Several of the brighter students had scored in the low nineties. An innocent word had booby trapped one of the multiple-choice questions, catching most of the college boys, and almost all of the essayists had failed to perform the sniff test to determine whether the candidate soil sample was loaded with forbidden peat moss. But Sergeant Pickett had drawn a perfect circle around every correct letter in the multiple-choice phase of the exam, and his essay, carefully printed with what must have been an extremely sharp 6H pencil (and therefore annoyingly difficult to read), contained all of the required tests and a succinct description of their purpose. I was about to red pencil a score of fifteen next to his answer when it occurred to me that Pickett would then obtain the only perfect score in the class. I hesitated. Should anyone receive a perfect score, on this or any other department of life? Surely if Williams could not fail absolutely, neither could Pickett succeed absolutely. The ancients knew that no one is ever self-sufficient, there is sure to be something missing. Man must be content to live within the earthly confines of two digits; the mere assignment of a score of one hundred was a dangerous arrogance, the creation of a monstrous mutation. I penciled in a fourteen on Pickett's essay question, a ninety-nine on the first page of his exam.

When Pickett heard his score, he requested an appointment to see me. We met in my office, and he asked me what additional commentary would have gained him a score of fifteen on the essay question. Unprepared to answer that one, I told him that no one had received a perfect score on the essay question, that, uh, there were many reasons why the tests should be run, beyond those which he had listed, it was, er, a matter of the completeness of his answer. I was afraid that he would press me further. But Pickett merely fixed me in his clear gaze for a moment and then said simply, "I see." Then he left my office. I never saw him again. But I wondered what he meant by "I see." Then suddenly I saw it, the tiny scratch, and exulted in my triumph. He should have said, "I see, sir."

1963   Unselfishness

To be a true leader, you must share the same dangers, hardships, and discomforts as your men.

The 588th Engineer Battalion (Construction) made its home on a pleasant wooded hill next to the Potomac. The Russians had just backed off in Cuba, and now in the sunny days of early winter the midnight alerts and rumors were over, and the 588th nestled comfortably back into its World War II–style barracks. I was leader of a heavy equipment unit, the third platoon of Company C, and in mid-December my men invited me to what they called a Christmas party. Conventional wisdom held that officers should not party with their men, but with two months left in the army, I was "getting short" and reasoned, as we are inclined to do when we are young, that conventional wisdom did not apply to my particular circumstances.

My platoon was a gang of roughneck equipment operators who spent most of their days down at the barbed-wire-enclosed motor pool, stealing parts from other platoons or sitting around the warm stove in their supply shack engaged in conversation that always stopped whenever I happened by. Tipton, my sergeant, prudently kept away from the platoon as much as possible. I soon learned that the man who really ran things in the platoon was Corporal Morton, formerly Sergeant Morton, busted three grades the year before for what some described as covering up for his men and others, more simply, as theft. Whatever the truth, Morton was a genial, fast-talking scoundrel who had mastered the art of making men continuously anxious and even competing among themselves to please him. I liked Morton, and the two of us quickly developed an unwritten contract in which my part was to protect him when his inclination toward recklessness got the upper hand and also permit him to deal with various indiscretions of "the boys" that might otherwise have been covered by provisions of the Uniform Code of Military Justice. In turn he ran the platoon for me and, what is more, assured me that the thousands of tools and parts that I had signed for without inventory would be in that shack when I left the army in February.

Morton himself did not issue the invitation to the party—that would not have fit the terms of our deal—but one afternoon when we were driving back from the motor pool, he suddenly asked me if I had been invited, and when I told him that I planned to attend the affair, he merely grunted and said that he would be there. I took this to mean that he approved and that he would be on hand to protect my position. "You know, lieutenant, none of that would have ever happened if I had got there when I planned to get there," Morton would tell me later. As it was, some domestic uproar detained him and he did not arrive until after midnight, by which time my troubles were well advanced.

Fresh from college fraternity life, I had naively expected the men of my platoon to be costumed in chinos and Harris tweed jackets, singing "99 Bottles of Beer on the Wall" in a setting festive with hanging crepe and the Kingston Trio. So there was a little letdown when I arrived at that minimally furnished, walkup apartment in a seedy part of Alexandria, to find no more than a dozen of the boys in T-shirts and greasy jeans, along with four of the toughest women I had ever seen, an ambience of glaring fluorescent lights, and cigarettes ground out on the living room carpet, with not a Christmas decoration in sight. I stood around awkwardly for a few minutes. After an initial glum greeting, "Hey, it's the lieutenant," no one spoke to me. And foolishly, oh so foolishly, I decided to anesthetize my discomfort with the vodka-Schlitz combo that appeared to be the featured beverage of the evening. Spec 5 Frazier, who was awaiting court-martial on a charge of stealing two Bailey Bridge jacks, mixed them for me with a sort of malevolent glee that was the closest thing I could find to cordiality.

After a while I noticed that no one was speaking to Willie Hooper either. Hooper was a forty-year-old black private, the only man in the 588th who knew how to assemble the rock crusher. Hooper had cleverly wheedled an official diagnosis of diabetic tendencies from some young army doctor, which provided him with a cover for drunkenness, which in his case was normally well advanced by noon each day. "We keep that boy around here for just one thing," Morton explained to me once, and sure enough, the week Khrushchev's missiles steamed toward Cuba and the crusher had to be assembled along with everything else in the army, Morton had locked Private Hooper in the parts shed and fed him coffee until he sobered up. So it came to pass that if in those fateful hours, President Kennedy had but given the word, the 588th could have delivered thirty tons of freshly crushed Virginia traprock to the White House lawn within the day.

By this time I was a little drunk and proceeded in my Boston way to deliver a fine abolitionist monologue to Hooper. Hooper, buoyed by my goodwill, passed some pleasantry to one of the surly blondes, who moments before had removed both of Spec 5 Frazier's hands from within the waistband of her jeans. In an instant, Hooper and Frazier became engaged in a quietly murderous dialogue. Hooper, much outnumbered and a man of basic goodwill, was trying to back out, but Frazier was systematically cutting all routes of honorable retreat. Observing this, I stepped up to the two men, unsteadily telling Frazier to go back to the living room ("That's an order, Frazier") and ordering Hooper to get his coat and follow me; I was going to bring him back to base. I was in that initial enjoyable stage of intoxication that women used to describe as "bubbly." It was all rather fun, the closest thing to combat I ever saw in the army. My finest hour.

The evening would have gone on to a happy ending if I had not driven through that stop sign in downtown Alexandria. Even then a simple signature on the notice to appear that the very pleasant city of Alexandria patrolman presented would have saved it. But by that time the vodka had disconnected whatever lobe attends to reasoned discourse, while leaving oddly unimpaired my powers of speech and locomotion. I became argumentative and then belligerent and fifteen minutes later was escorted into a jail cell at the Alexandria police station. (Hooper meanwhile had slipped into the night like a phantom.) From time to time the desk sergeant and then later the two MPs who were called in to provide consultation explained with kindly persistence that I did not need to spend the night in jail, if only I would cooperate. One of the MPs held a clipboard on which was mounted a pad of standard Department of Defense forms meant for the taxonomic classification of various disorderly members of the armed forces. The form included various boxes to check designating the subject as "neatly attired," "in uniform," or "disheveled," "cooperative" or "disorderly." I foolishly insisted that as the ranking member of our group I should fill out the form, then proceeded to amuse myself by checking the least favorable descriptor in each category. It was an inconvenient manner of self-evaluation, but there was perhaps an element of some honest confession in my madness. We Irish are subject to spells of self-criticism.

The Alexandria police were splendid. They gave me my own cell, and they even called my wife and cheerfully explained that it was a small matter, that thirteen dollars bail would do. Meanwhile I sat for an hour or so alone in the cell. It seemed funny at first, then depressing. The bars. I could not sleep.

Memorable sounds, footsteps in the corridor, the clank of the cage door. "Your wife is here to take you home," they said. The laundry money, our savings, stacks of quarters and dimes in a row on the desk. My infant daughter in a pink acrylic pajama blanket, mounted in her plastic carrier, gurgling happily at the kindly desk sergeant's tickle.

I stood disgraced and surly, holding up my pants. Then I put on my shoelaces and belt, and my wife drove me home.

Later that morning at the Company C orderly room, I fought nausea and remorse with coffee and action and within four phone calls had reached and identified myself to Major Edwards, military police, Washington, D.C.

"Well, well, yes, lieutenant; we have an interesting and I must say unusual report on you this morning," said the major. My tour of active duty was almost finished and anything he could do, to, ah, minimize serious consequences, I'd appreciate that. Well, he'd see what he could do, said Major Edwards.

The order to appear before Colonel Bickerman, battalion commanding officer, came down a week later, on a Saturday morning. I saluted formally, "Lieutenant Meehan reporting as ordered, sir," a quiver in my voice. Bickerman left me standing at attention. A grave sign.

"It's bad enough when I receive this kind of report on my enlisted men," he said, scowling at the form on his desk. At the bottom of the paper I could see a large, page-wide box that I suspected was entitled "General Remarks." It contained four typed lines. How had they squeezed so egregious an offense into such a short paragraph?

Bickerman looked out the window. I heard, still hear, the distant shouting chorus of men at drill, the staccato beat of a helicopter, Bickerman speaking in a tired quiet voice, the prologue to my sentence.

Have times changed? Do men even into their twenties still feel that primeval awe as they stand, like little boys with doe-soft eyes and hairless bodies, before fathers, priests, headmasters, commanding officers, and bosses? Surely they do, for it is only under such circumstances that a man, at age twenty-four, can discover the carefully guarded secret of the tribe: that the thunderer before him is only another little boy.

Colonel Bickerman looked up. He wore plastic-rimmed glasses, there was a redness around his nose and cheeks, and his grey eyes were watery and tired. He had a hangover. "I expect more from my officers than this kind of behavior. You yourself have been involved in my campaign to improve traffic safety in this unit. It hardly reflects well on us to have the officers of this battalion stopped by civilian police for running stop signs. I don't want to see any more reports like this on you while you're under my command. That will be all."

My heart flooded with joy and gratitude. Suddenly I saw for the first time that all along, everyone, Morton and the Alexandria police and Major Edwards and now Bickerman himself, had tried to protect me, for those two gold bars that I polished with Brasso each night. It mattered little to them that I had considered myself a costumed fake. "Sure, the lieutenant has his faults; so don't we all," they thought. But still, he *is* an officer. Soon I would be saying good-bye to all that. But now, just before the end, I was grateful to them all.

There were two dozen houses in the village of Bu Hua Chan, each raised on hardwood stilts in the Thai fashion and sheltered from the tropical sun by tall coconut palms. This shady grove occupied a terrace on the left bank of the Lam Pra Plerng, at a point where its brown waters ran out of the hills to wander across the featureless terrain of the Korat Plateau. In the cool dawn, roosters crowed (in Thai, ecky-ecky-ay) and the village monks, with their clean, spicy smell and robes the color of saffron, walked with silent gravity on the muddy path from the monastery into the village, carrying dark clay bowls and accepting without sign of acknowledgment the boiled rice offered by old women at each household. Later in the morning as I drove my green jeep down the rough laterite road from the village, little boys stood on the backs of savage-looking but cowardly water buffalo, waving and smiling at me. In the heat of day, young women bathed in the river, wet hair jet black against caramel skin, and old men retrieved miraculous nets of wriggling fish from puddles that formed during the monsoon in every mud-cracked hollow in the land. Later in the evening there would always be a cool bottle of Singha beer for me at the house of the headman or the abbot or the schoolmaster, and sometimes my companion Vicha and I would sit talking with them of the history of the village, and of the spirits, friendly and unfriendly, that flitted through the forest and peered at us from the shadows.

In other respects and at other times, Bu Hua Chan was less picturesque. There were insects that laid eggs in your skin as you slept, and I remember one morning almost fainting at the puff of humid fecal air that enveloped me when I entered a Gauguinesque cluster of banana trees that served, with admirable ecological economy, as community toilet and fruit garden. Vicha once told me that the former assistant headman, Longmah's predecessor, had split another man's skull with a machete fighting over a woman, so I cannot say that the village was a social Garden of Eden. But on the whole, it was not a bad place, for there were "rice in the fields and fish in the waters," as the Thai defined the good life, and I cannot say that people are any happier in the suburban neighborhood where I now live than they were in Bu Hua Chan.

I suppose there is no trace of the village today, only a great brushy hole. For later we came down on Bu Hua Chan with a fleet of yellow scrapers and Caterpillar D8s, and took the village for the twenty feet of silt beneath it for the dam. By that time, when we finally moved the houses and brought out the monks for their prayers so that we could begin grubbing and ripping up the burial ground, I was sick of the tropics; I was bored and lonely, and with the Thai whiskey and diarrhea and heat, I was beginning to believe what a Dutchman from Jakarta once told me, that I would not live to see forty-five if I stayed in the Far East.

Like many of the best and worst adventures, my assignment in Thailand was not planned. One summer afternoon it just happened. I was twenty-four years old, just out of the army, and had taken a job in New York City, working as a dam designer for a big architect-engineer firm that specialized in foreign work. I had the idea that they might send me overseas, to some exotic place, for I had a New Englander's fancy for warm lush places in coral seas, for rum and coconuts, dark eyes and bare breasts, secret ruins overgrown with creepers and hibiscus and populated with wicked monkeys and bright parakeets. As anyone who has suffered from romantic fevers well knows, the pleasures of distant lands are much enhanced when they are also a century distant in time. Like Miniver Cheevy, I scorned the present and yearned for the days of old. Unmoved by the waves of excitement emanating from Cape Canaveral, I resonated to engineered glories of bygone days, the Crystal Palace, Brooklyn Bridge, Hoover Dam. I heard the echo of Walt Whitman "singing the strong light works of engineers." Puffing at my pipe, I surveyed our handiwork, Panama and Suez,

Where the tribes of man are led toward peace
By the prophet-engineer.

Alas, my employer, the firm of Tippetts-Abbett-McCarthy-Stratton, believed that I would be more useful running Atterberg Limits soil tests on Park Avenue than giving gifts of T-shirts and Zippos to natives in some distant rain forest, so I remained in the grubby reality of New York, the apartment in New Rochelle instead of the grass hut in New Caledonia. Disgruntled, I took to watching the late show and drinking cheap beer. In the growing heat of morning, I fermented on the commuter local to Grand Central, reading the *New York Times* on stale straw seats. At work I had the responsibility for studying soil samples gathered at dam sites on remote stretches of exotic rivers. Crates of black cotton clays and pink lavas would arrive almost every week from the creeping jungles of Columbia, the bandit-infested hills of West Pakistan, the shimmering plains of the lower Ganges, the green-terraced hills of Taiwan. Condemned to an air-conditioned glass box, I squeezed whatever vicarious pleasures I could from the field notes of men who were living the adventurous life that was supposed to be mine. And yet the firm had some challenging assignments, and the work fired my imagination. The Nile was backing up, threatening to flood the temples of Abu Simbel. Could I stop the rising waters with a cofferdam long enough for the contractor to saw up the temple and cart it off to a new location? Could a scheme be devised to control potentially fatal leakage beneath the mammoth Tarbela Dam? I sketched and plotted, calculated and designed, but I knew my solutions were abstract, naive. I needed experience in the field, knowledge of real construction techniques, of mud and monsoons and malaria.

Salvation came to me one Friday evening in the form of a long-distance phone call from Denver. It was the vice-president of another engineering firm. He had seen my resume. Was I interested in a two-year assignment overseas, in Thailand? My heart skipped a beat. Thailand! The king of Siam! The river Kwai! What more could I ask for? I would be an adviser to the Royal Thai Irrigation Department, the vice-president said, building a dam in the far up-country. They would send me airplane tickets. I could fly either way, east or west. The fare was the same either way.

Suddenly it had fallen into my lap—a free ride to as far away as anyone could possibly go. Beyond New York, across the wide Missouri, California, to the end of the continent, over the South Pacific, past Iwo Jima, Shanghai, to Thailand. To the end of the trail. Naturally I accepted.

Bangkok, Thailand's capital, is sited on the delta of the country's principal river, the Chao Phraya. In the beginning, they are all the same, these river-mouth cities, Babylon and Alexandria, New Orleans and Calcutta, reedy swamps, the horizon a hazy unbroken line. Below that line, the yellow greens and murky browns of a silt-filled sea. Above it the vast dome of the sky, at times a panorama of cumulus clouds, but more often, in my mind's eye, domain of a tiny savage sun. No city planner or municipal engineer would ever choose to build a city among these wandering tidal sloughs and quaking islands of mud and rotting vegetation. Here sewers choke themselves and spill into the water system. The Buddhist pagodas tilt grotesquely in the bottomless ooze, and every building more than two stories high must be carried on wooden piles brought from distant forests and driven deep into the muck. Roads, if they are to remain firm during the rains, must be laboriously banked up. Navigable waterways, arteries to the heart of commerce, suffer mud-choked coronaries. The inhabitants, farmers who once offered prayers of thanksgiving toward the annual return of silty floods, give way to a middle class that considers muddy water an imposition. The warm haze, rich with the odors of life, becomes ochreous with sour cooking smells and diesel fumes.

Perhaps it would not be fair to say that Bangkok was entirely lacking in touristic charm. I was told that the floating market and the one-day tour of the city's canal thoroughfares were interesting, although I have not been to them, and I can give eyewitness testimony that the city featured several house-sized plaster statues of the Buddha, decorated with impressive quantities of broken bottle glass. For the lazy or overwrought—and most foreigners in Bangkok seemed to be one or the other—a standard ten dollar bath was provided gracefully by an attractive woman, with a full course of other delights available at nominal extra cost. There were dozens of nightclubs, and the Filipino musicians who entertain in them today presumably do numbers by the Bee Gees that sound more like the Bee Gees than the Bee Gees do themselves, just as they did with the Beatles when I was there in 1963.

On September 21 of that year, Pan American's flight number 2 descended through ragged monsoon clouds to the green, weedy sea of the lower Chao Phraya plain. In the terminal building of the perpetually unfinished Don Muang International Airport, Thai customs officials, chattering among themselves in a curious singsong way, indifferently checked the luggage of the dozen or so American passengers. Most of them were military advisers who had come to train the Thai military in techniques of counter-insurgency. Within minutes, taxi drivers were carrying the passengers' suitcases in a minced run to miniature Japanese cars with plastic-covered seats. My cab was a Toyota, the first I had ever seen.

We sped from the airport through paddy fields toward the capitol. Around us the sun flashed through the broken clouds on fluorescent yellow-green young rice shoots; in the distance, soft gray slanting columns of rain moved silently across the plain, wet gusts of warm wind rippling the fields around them. I had come from New England, where the line storms lingered and brooded for days or weeks, the sky an aquatint backdrop that remained long enough to provide a scene to important passages of life; I remember that my grandmother's sickness, death, wake, and burial, a cycle of events that lasts a week or more, took place in the same soft gray drizzle. Here, closer to the equator, the weather was more vigorously adolescent. A rainstorm could be framed in a few minutes' time and in the space of a few miles. First, a gusty wind rippled the fields. Then with a roar, sheets of rain enveloped everything; water spouted from rooftops and ponded on the streets and lawns. Like an attack of momentary rage, the storm passed, and it was strange to see the dazzling tropical sun flashing on the rivulets of still-moving rain water and misting the wiry Bermuda grass lawns.

The company put me up for the first week after my arrival in the Vieng Tai Hotel, a Chinese-owned, pastel plaster and pink-tiled hotel of seven stories in the commercial section of the city. The Vieng Tai was set on a shop-lined street, which, for the eighteen months that I spent in Thailand, was continuously (and perhaps is to this day) torn up for the installation or repair of some underground pipe. It was a middle-class hotel that catered to Japanese salesmen, up-country Thai and Chinese businessmen, and

middle-level foreign advisers and technicians like myself.
I began my days there with fresh-squeezed orange juice
and cold fried eggs served with cheerfully inept formality
by teenage Chinese waitresses. Mornings were spent at the
Royal Thai Irrigation Department's fortress-like com-
pound, meeting Thai officials, participating passively in a
mysterious series of currency exchanges, passport stamp-
ings, and paper signings, which completed my metamor-
phosis from tourist to temporary resident status, and
eagerly attempting to sketch in a broad outline of my responsibilities on my up-country project. In the afternoon,
I crisscrossed the city in taxis and three-wheeled samlors,
shopping for canned goods, pots and pans, and Thai gram-
mar books, in preparation for my move to the Northeast. I
shucked my wardrobe so carefully planned back home
and outfitted myself in the expatriate's standard tropical
uniform: dark blue trousers and white short-sleeved shirt,
custom made by a bearded Indian tailor named—like all
other Sikhs, I was to learn—Mr. Singh. I began to practice
the tonal inflections that add an essential dimension to the
Thai language. I learned what every foreigner claims is the
favorite expression of the Thai, *mai pen rai*, which means
"It doesn't matter," and is thought to reflect the casual Thai
attitude toward life. Of course the Thai do not look at it
that way. "We are always saying that to foreigners because
they are too nervous," one of them told me. "The way some
foreigners walk, we say it seems they are always chasing
their water buffaloes," another said.

Bangkok lies in the heart of Thailand's geographical rice
bowl on the banks of the Chao Phraya River. The Chao
Phraya rises in the mountainous northern part of the country and flows due south to discharge its muddy waters into
the Gulf of Siam; its flat floodplain divides the country
down its middle. To the west of its lowland paddy fields
are the rugged hills of the Bilauktaung range, cleft by the

River Kwai, a country that I do not know. Twenty or so miles east of the river, the monotonous flatness of the floodplain is interrupted by a belt of jumbled limestone hills, picturesque and jungle-covered, useless and unfriendly terrain from the Thai point of view. The hills mark the scarp boundary of the Korat plateau, a flat slab of terrain the size of Texas that some forgotten geologic paroxysm has lifted five hundred feet above sea level. The plateau is tilted toward Laos to the northeast and drains into the upper Mekong River, which divides Thailand from Laos.

So it is that when one travels by highway from Bangkok to the provincial city of Korat, 130 miles to the northeast, one leaves the fertile lower central valley two hours out of Bangkok, winds and climbs through jungle-covered mountains for an hour or so, then emerges into what, from a geographer's point of view, should be another country— scrubby, rolling, dusty terrain supporting widely scattered villages and patches of dry-farmed rice. Here, within a few tens of miles from the Thai capitol, one is in the upper reaches of the Mekong basin, and although the brown waters of the great Mekong River still lie hundreds of miles to the east, that distant river exerts a pull on the culture within its basin that is as strong as its attraction to every wash, creek, and river that cuts the dusty surface of the plateau. For just as the geographers tell us, the flow of culture follows closely the flow of water.

Here on the plateau the people take on the lean character of an upper watershed Asian peasantry. They walk on dry cracked feet, heads swathed in turban-like coils of the universal cotton *pacoma*, wearing loose black cotton jackets and short pants, tending scrawny chickens and attended by scabrous dogs. Here also the yield of the land is thinner; bits of charcoal made from brushy trees that may not grow back, spindly rice crops that after a few seasons of stubble burning peter out, leaving a soil sucked dry of nutrients and well on its way toward becoming a permanent desert. I am not a believer in the a priori virtues of virgin wilderness, for I have seen landscapes where the hand of man has done fine things. But here on the Korat plateau, there seemed basis to the argument, pressed by a visiting UNESCO delegation in 1959, that an irreversible process of exhaustion was at work. "Drought in the Northeast will gradually increase, and more areas will turn into semi-desert," their report warned. Man, even without the aid of modern technology (save, perhaps DDT, which had stimulated population growth by controlling the malaria-bearing anopheles mosquito) had become a pestilence on the land.

Since the twelfth century, when the Chao Phraya basin people began to emerge as a national identity and wrested the Korat plateau from the Cambodians to the east, the Thai have held political control over the mixed peoples of the Northeast, the Lao, Cambodians, various hill tribes, refugees from the several postwar Indochina conflicts, including most recently those fleeing the Cambodian genocide. One might consider the Northeast to be a sort of Thai colony, a buffer zone held not so much for its economic value but as a shield against the barbarianism, foreign control, and a recent years, communist dominance of the Mekong basin. The strategic significance of that buffer has not been lost to the United States. Following World War II, we financially supported Thai efforts to forge strong economic and communications links with the Northeast, to penetrate that mountainous divide and counterbalance the dangerous, ever-present geographic draw of the Mekong. Then in the late 1950s, AID built Friendship Highway; more recently, the U.S. military built a highway bypass that links the town of Korat, gateway to the northeast, directly with the Gulf of Siam. For the past two decades, these have served as the arteries through which agricultural assistance, medical teams, military ordnance, special education and police missions, and other assorted cultural paraphernalia have been pumped to maintain this grafted lobe of Thailand. Even then regional economic indicators of the early 1960s showed that the Northeast, with an economic growth rate of only 2 percent and a per-capita income of seventy dollars, lagged badly behind the rest of the country.

In 1961 when the Thai government's Central Development Board set forth its five-year, $300-million plan for national economic development, a program that was to be financed in large part by U.S. aid, the development in the Northeast of modern, irrigated agriculture as an alternative to the slash-and-burn crop practices that had been earlier criticized by foreign experts was assigned high priority. Of the five medium-sized irrigation projects scheduled for development between 1961 and 1965, two of them, the Lam Pao and Lam Pra Plerng projects, were in the Northeast. It was to the latter that I was to be assigned as an adviser.

The Lam Pra Plerng River rises one hundred miles northeast of Bangkok in the rugged belt of wet forested terrain that forms the divide between the Chao Phraya floodplains and the Korat plateau. The river flows for about five miles toward the northeast in a pretty, mile-wide, jungle-covered, uninhabited valley flanked by creeper-draped sandstone bluffs; then it discharges onto and wanders indecisively over the flat terrain of the plateau, toward the town of Korat. At the end of the monsoon in October, the river gushes 25,000 cubic feet per second of silty brown water, spilling its banks and flooding the plain for a week or two. But for much of the year, especially toward the end of the dry season, a brush-tangled stream bed dries to a series of slimy pools that reek of buffalo urine and breed mosquitoes. In this condition the river is neither attractive as a source of drinking or bathing water for the scattered palm-covered villages that one finds each mile or two along its banks nor sufficient in capacity for dry-season irrigation of the floodplain paddy lands that flank it for a mile or so on either side.

A couple of miles back into the hills is a narrow stretch of the river where the canyon is as wide and deep as a broad city street flanked by ten-story buildings. This site had been considered for many years by Thai engineers and endorsed by a team of U.S. Bureau of Reclamation advisers who had been sent to Thailand in the late 1950s, as a possible site for an earth dam. Here, with some effort, enough earth—roughly 1.5 million cubic meters, 300,000 dump-truck-loads—could be scraped from nearby hills and thrown up into an embankment that would block the canyon, backing a narrow reservoir several miles up into the valley. The reservoir would capture much of the monsoon floodwaters, which could be held until the dry season and then released gradually to permit artificial irrigation of either the existing paddy lands (thus permitting a dry-season rice crop) or additional cash crops on the fallow, brushy lands outside the floodplain. Construction of the project required first that an earth dam be built, complete with a valved concrete pipe outlet works through which controlled release of the water could be made, and a spillway structure that would safely discharge past the dam any floodwaters that entered the reservoir when it was full. In addition to the dam and associated headworks, the project required the excavation of an extensive network of canals that would convey water to the irrigated lands below.

By mid-1963 when I arrived in Bangkok, a feasibility report had been prepared by the consulting firm that had employed me—a Denver firm started by the same USBR engineers who had been advising the Thai Irrigation Department in the late 1950s—and the project had been approved on that basis for an AID loan in the value of $7 million. As customary, a condition was that the portion of the loan provided in dollars be spent on U.S. goods and services, including retention of U.S. consulting engineers to carry out the necessary designs and purchase of a fleet of construction equipment, bright yellow Caterpillar and International Harvester tractor bulldozers and earth-moving scrapers, necessary to complete the construction. By September 1963, funding of the project had been approved, and design was underway at the Denver offices of the consultants. At the site, an access road had been hacked through to the brushy plateau overlooking the dam site, and a construction headquarters was being established.

Each of the AID projects of this type required a soil engineer whose job was to set up a soil-testing laboratory and direct whatever exploratory excavations on the site were necessary to reveal the character of earth and rock materials that existed in the subsurface, essential information for the design engineers to plan the final layout and type of structures. Later when construction was underway, the soil engineer's job became to check subsurface conditions as they were revealed in excavations and to test fill material being placed in the dam embankment to ensure that the earth was being compacted adequately. The American engineer was to act in an advisory capacity, with responsibility for action being held by a Thai counterpart engineer and his staff of technicians. It was to this position, soil engineer on the Lam Pra Plerng project, that I was assigned.

I stayed at the Vieng Tai Hotel in Bangkok for ten days, clearing my household goods through Thai customs and meeting and being entertained by Thai officials and the local staff of the consulting firm. These Americans were a cheerful crew and hospitable, mostly expatriates with years of overseas experience. They seemed like leathery Hemingway characters to me, and I was impressed with their worldly outlook and their humor laced with a touch of cynicism. Some of them had left the assorted debris of failure behind them in the States and were making a career of overseas work. The pay was good, and Bangkok was a pleasant station. Up-country assignments were a different matter. A rough life. There was the story of poor Ferguson, the soil engineer on the Lam Pao project. He started sleeping with his cook, bad form in Thailand. Drank too much and ran around with a pistol. One day on the train to Bangkok he flew into a rage over some seating dispute and screamed at the first-class passengers, "Your king is a one-eyed son of a bitch." The company had to let him go after that one.

But I disapproved of the Bangkok group too, with their politicking over post exchange privileges, the gleeful contempt they sometimes had for the Thai, their stock market investments and subscriptions to *Time*, the ignorance they had of the local language. "Thai food?" one of the wives said to me with a smile. "I think of it as sort of warmed-over garbage." On the job, I noted those little cross-cultural tensions that developed between Thai and Americans. The Thai engineers were lazy. The Thai engineers did not show up at meetings. The Thai engineers said yes when they meant no. The complaints were more than a century old. America's first ambassador to Thailand, a New England trader named Townsend Harris, retired in disgust from the country in 1856, remarking, "I hope this is the end of my trouble with this false, base and cowardly people; to lie here is the rule from Kings downward."

I have a diary from those days, a local product with a plastic cover, an almanac of royal history, lined pages with dates in both Thai (year 2506) and English (year 1963), and a frayed yellow ribbon to mark my place. I paid $1.50 for it and remember clearly my concern with its apparently cheap construction. I worried that its spine might crack. But the diary remains in good condition today, and my anxiety in retrospect appears to be a confusion of form and content. Its first entry, Saturday, 21 September, is "Arrived Bangkok Airport 11:30 A.M. Settled at Vieng Tai Hotel." The next few days are filled with inspections, briefings, meetings with this and that official. By Tuesday, 1 October, I am impatient with all that, especially with my fellow Americans and their complaints. I am ready to head out to the jungle, to know the villagers, to talk religion with the monks, to conquer the Lam Pra Plerng with my dam. On Wednesday, 2 October, I leave Bangkok by Land Rover at 0800, arriving Korat 1200. It is the end of the rainy season, and the road to Lam Pra Plerng is flooded. But on Friday, 4 October, I make my way through to the dam site. There I am surprised. "Neither the project manager, assistant project manager, or resident geologist were present at site—all in Bangkok," my diary observes, in an offended ambassadorial tone. In a small way, for the first time, Thailand has let me down.

But in other respects, Lam Pra Plerng was all that I had hoped for. The rainy season was coming to an end, and work on the construction camp was just getting underway. Down on the river, which was flowing heavy and silty brown, gangs of sweating laborers hacked at juicy stalks of vegetation with machetes, driving back the jungle. Elephants skidded hardwood logs from the forest, dragging them with chains onto the plateau overlooking the river. The carpenters cut them there and erected building frames, which they roofed with palm leaves. The camp sang with this labor and steamed under the fierce midday sun. Late in the afternoon, gray clouds from the Gulf of Siam crowded the sky, and white sheets of rain pounded the forest and the metal roof of our new office building. Then one day the rain did not come, and my driver, Luay, said the monsoon was finished. As the axle-deep mud in the camp began to dry and stiffen, we dumped red laterite gravel on the roads, and each day new vehicles, rattling olive-drab jeeps and gray Land Rovers, and then scarred bulldozers and big yellow scrapers, made their way to our camp and settled there.

I spent my days at the dam site planning the program of subsurface geologic exploration necessary for the design of the dam and its various works. At first we had no equipment suitable for this work, but then one day a drilling machine, an antique from the pages of an old mining journal, mysteriously appeared at the camp (equipment appearances and disappearances were always mysterious), and we began to drill core holes in the yellow sandstone walls of the canyon, testing their soundness and watertightness. I hired crews of laborers from Bu Hua Chan and taught them the arts of digging test shafts and drilling auger holes. I got to know some of the local people and met the assistant headman of the village, a farmer named Longmah.

During the eighteen months that I spent in Thailand, I developed close working relations with perhaps a dozen Thai at various levels within the Royal Thai Irrigation Department. This was relatively easy for me to do because the Thai are a curious people, and I was the only American on the project. I had achieved an instantly prestigious position in the camp when Pira, the project manager, a squat, scowling, authoritative Thai who was greatly feared by the fifteen hundred engineers, equipment operators, foremen, and laborers who worked at Lam Pra Plerng, graciously assigned his driver, Luay, and his canvas-covered jeep to my exclusive service, leaving himself an old broken-down jeep of World War II vintage. Whether this transfer was Pira's idea or whether it had been ordered by the Irrigation Department's director general, I do not know, but like many gestures that are made at the outset of such a relationship, when the value of sacrifice is better understood by the donor than the recipient, the gift in the end became a source of friction and did more to divide than cement us.

Nonetheless if Pira's gesture did not ensure harmony between us, it provided me with four-wheel drive mobility, a mark of authority that perhaps transcended my actual significance on the project and guaranteed a courteous welcome in any village or town in the region.

In Bangkok at the Irrigation Department headquarters, where I regularly spent several days a month discussing field operations with the project designers, planning subsurface exploration programs, writing reports, stocking up on provisions, and going to nightclubs, I was recognized as the sole American representative on the job. At least in the early days, I was always treated with elaborate cordiality by the senior Thai engineers and officials. This was in part a matter of traditional Thai hospitality and open friendliness, which I believe was entirely sincere, but also in consequence of my role as an observer for AID, which was funding the project. The Thai, who participated in scandal with a guiltless enthusiasm that I envy to this day, were nonetheless well and wisely aware of the fussy attitude of many Americans toward the disposition of U.S. loan funds, and I believe they dreaded that one of my daily reports would suggest that they were using pirated U.S. funds to build teak-paneled vacation bungalows for the use of high Thai officials and their mistresses. But I soon learned to sanitize my daily and weekly reports of such suggestions, and after a discussion one day with a disturbed Pira, even stopped including parenthetical remarks pertaining to when he was and was not on the job.

My supervisor in Bangkok, Burns, urged me to send in frequent reports, preferably by radio, and I eventually achieved the sense of meaningless communicative urgency that was suggestive of great progress and made everyone happy. I can remember now, passing the project radio shack in the heavy quiet and fading heat of early evening, hearing the Thai radio operator, his face shining with perspiration in the light of a kerosene lantern, loudly struggling with the transmittal of some idiotic message that I had written that afternoon.

As I look back on it now, my principal role on the job was to be the American presence there, to satisfy everyone that our foreign aid was being put to good use. I did not understand this at the time, thinking that my sole responsibility was to explore the dam site and provide data and suggestions that would ensure a safe and economical design. In fact I deplored the politically motivated behavior of people like old Dodson, the retired Soil Conservation Service farm adviser who was supposed to be assisting the Thai in crop planning for irrigated lands. Dodson was assigned to Bangkok, but he appeared from time to time on my project with a retinue of smiling officials and a photographer. At his instruction, they would find some cute schoolgirl in her blue and white uniform, carrying her little notebook. Dodson would set up some elaborate pose of himself fingering the black clods of soil or pointing down some canal, supposedly explaining something to the schoolgirl, who would stand with a wondrous expression on her face, gaping at the antics of this elderly foreigner. As Dodson well knew, the picture would show that she was impressed with what U.S. advice and assistance were going to do for her future. Of course Dodson did not even know how to ask for a men's room in Thai, and the picture as often as not would be set in some locale, nearer to Dodson's hotel in town, that had nothing to do with the AID project. I remember once that Dodson, with what was no doubt cagey satisfaction, did his public relations shots at an Israeli-run demonstration farm.

I still have a copy of one of our monthly progress reports to the AID. Its pages are yellowed and faded and still have that cool, permanently damp feeling that all of my books took on in the monsoon and that storage for the last fifteen years in California has not dispelled. The report includes a dozen or so glossy black and white photographs of our group of AID-sponsored projects. My work is reflected in a series of photographs that show excavation of the outlet works at the dam. Here is one of myself, posing with a sweaty important look, inside a giant crack in the rock of the left abutment of the dam. I was proud of that picture. It showed what a tricky adversary nature was proving to be and suggested that the project would be successfully executed only through the engineering efforts of clever fellows such as myself. A few pages farther on, Dodson, dead some ten years now, is memorialized, pointing a gnarled finger down the alignment of some weedy canal lateral, under the watchful eyes of that befuddled schoolgirl. I am now perhaps a little wiser, more cynical, and a dozen years removed from the stuffy puritanism and engineering arrogance of my youth that even now seeps into my recollection of Dodson's efforts, and I confess that Dodson's picture is much more interesting than my own.

Of the various roles we advisers were to take on in our assignments—data gatherers, designers' representatives, reporters—the one that brought us in closest contact with the Thai was the role of counterpart adviser. We were supposed to work on a one-to-one basis with a Thai official with similar training and background who held a position within the hierarchy of the Irrigation Department that corresponded to our own professional position. The idea was that we would provide continuous advice and consultation on technical and project matters, and the Thai official would actually carry out the operations that we suggested, presumably learning something of our American methods in the process. This system worked reasonably well because the Thai admired American technology and knowhow and were naturally friendly to Americans and usually receptive to advice. The higher-level Thai officials professed to admire American efficiency and encouraged the younger staff to compromise their easy-going, pleasure-seeking, people-oriented ways with the puritan work ethic. In those days, I took a grimly serious view of work and perhaps for that reason was considered a good adviser by the senior Thai officials, although I suspect that the staff people with whom I had daily contact on the job soon became uncomfortable with my attitude of stern disapproval of their erratic work hours. Bad weather and marginally passable road conditions frequently impeded operations on the project. At first my attitude was to storm and rage in the face of these environmental insults, and I took delight in making heroic attempts to struggle through torrential downpours, axle-deep mud, and impenetrable jungle thickets, whereas the Thai were inclined to shrug diffidently, invoke the Thai saying, "mai pen rai," and retire to a local café for an afternoon of leisurely drinking and genial conversation.

Whatever my Thai laboratory technicians may have thought of my grave admonitions about keeping regular hours, they made an effort to comply with them, if only to please me. One morning a few days after I had set down attendance rules with the severity of a Marine Corps lieutenant, I arrived in my office to find one of my local village recruits sprawled on the floor in front of my desk in a drunken stupor. When I disgustedly requested that he be removed from my presence, I got some very disapproving looks from the engineer who was with me. "You told him he should be at work every morning at nine," he said, "and last night at the festival he told his friends to bring him here for sure. He won't understand if you're angry at him."

The Thai knew that we advisers were being paid to give advice and out of politeness always pretended to be impressed by our wisdom even when the advice was impractical, gratuitous, or idiotic, as was sometimes the case. Those of us who were accustomed to the bristling arguments that are the rule on U.S. construction projects suddenly found ourselves in an environment where our words were always greeted with apparent courteous agreement; and in this rather different setting we sometimes failed to recognize that this reception and the assurance of compliance that it suggested was no more than an attempt to avoid hurting our feelings, and in some cases, a mask for much harsher judgments than we were aware of. I remember one conference at the office of our project manager, Pira, in which an American consultant, just arrived in Thailand, earnestly explained to us all that during the rainy season, soils in the areas that we planned to excavate for embankment fill would become subject to the action of rain water, which would result in an increase of their moisture content and a consequent softening that would make their trafficability by wheeled vehicles less favorable than during the dry season. Pira knew, as his father and forefathers had well known before him, that things got muddy in the monsoon, but nonetheless nodded with seeming grave respect at this information. But I saw somewhere in the back of his eyes a contemptuous click that signified that this man was a fool, and I knew that nothing else that the consultant said—and he went on to say a great deal more—would be listened to seriously by the Thai. Nonetheless the Thai, who placed a high value on saving face, made sure that he went away from the meeting feeling that he had made a major contribution to the project.

My Thai counterpart engineer during the first few months of my stay at Lam Pra Plerng was a young Thai engineer named Vicha. His father was the governor of the province in which our project was located, and he lived in a rambling, musty, old wooden house that served as the governor's mansion, located on the edge of the town of Korat next to one of those striking orange and white Buddhist temples that the Thai built to gain merit but then never visited. The family was of distinguished lineage, from the central plains. The governorship was evidently a rotating post filled by appointment by the minister of interior, for his father had served previous terms as governor of other provinces in the northern part of the country.

Through Vicha I met the governor several times at his home and at various social functions to which I was invited. He was a short, gentle, round-faced man, exceedingly polite and gracious, who made little of his powers and responsibilities as the central government's nominal representative to the sensitive province of Nakorn Ratchasima, gateway to the Northeast. This may have been due in part to modesty but perhaps also reflected the limited real powers that the civilian representative of the Thai government actually exercised. It was always my impression that the provincial police chief, who reported directly to Bangkok, wielded more power than the governor.

Nonetheless the governorship carried a great deal of social prestige. Vicha told me, with much amusement, of an episode that had occurred a few months previously and had been widely covered in the sensationalistic Thai press. The newspapers reported that the governor practically single-handedly had captured a group of ferocious bandits in some remote northeast town, following a fierce gun battle that had lasted for several hours. Of course the governor had never left his headquarters, and the bandits were more likely forlorn ragamuffins, but the romantic version served both the public appetite for entertainment and the governor's need to play a heroic role, and everyone agreed that the facts of the matter could be allowed to pass the board. I observed more than once that the Thai had a distinct preference for a good story over a lot of dull data. "Thai custom," Vicha would tell me, smiling with pride and approval, when I would remark on this fanciful approach to life.

At first I lived in the town of Korat and commuted daily by jeep sixty kilometers to Lam Pra Plerng. Each morning at eight o'clock, Luay, my driver, and I would pick up Vicha at the governor's mansion on the edge of town. Invariably various petitioners, mostly poor women with children in arms, would be sitting on the front steps, waiting to present some cause to the governor. Vicha would exchange a few princely words with them, and then the three of us would drive out of town to the rough, unpaved road that led to the mountainous terrain to the south. The Irrigation Department had a local branch office and radio communications center on the edge of town, and we would frequently stop there to send messages to or receive them from their headquarters in Bangkok. Then weather conditions permitting, we would drive through the plains south of Korat, through the silk-weaving village of Pak Tong Chai, to the point where the new canal intersected the highway. Then we would turn onto the new laterite-surfaced road that took us into the hills to the project headworks and dam site.

Vicha and I spent most of our days together during my first months in Thailand, and my recollection of the working relationship and friendship we shared is strong. For both of us, it was a first exposure to another person of similar age, education, and even personality but entirely different racial and cultural background. Both of us had an instinctive curiosity about the extent and depth of these differences.

Vicha was a big man relative to most other Thai, but he carried himself with the grace of a dancer, and the softness of his voice and delicacy of his gestures suggested, to a Westerner, a certain effeminacy, although it was my impression that to the Thai it was taken as a mark of good breeding. He took it as his responsibility to introduce me to Thai culture, and as we traveled through the rural areas near Lam Pra Plerng, seeking out sources of various construction materials, we met and talked with farmers, shop owners, monks, village headmen, hunters, and government officials. Vicha would approach them all with elaborate and earnest courtesy, listening intently to their views, smiling sympathetically. Like the Boston Irish politicians I remembered from my youth, he was able to draw people out and make them feel significant and interesting. They would invite us for tea or beer in the towns, and in the backcountry to go hunting or to visit their stilt-supported houses for a snack of hot peppers and salt.

Most of the Thai engineers wore the standard khaki uniforms of the Royal Thai Irrigation Department, but Vicha, always the aristocrat, preferred rough silk shirts and trousers and slipper-like shoes that were almost a comic contrast to my heavy GI boots. In true American style, I would slog through muddy jungle trails; Vicha would walk beside me, delicately bypassing various obstacles and hazards.

We covered a lot of ground during those first few months, and under one pretense or other, we managed to visit most of the villages and towns in the region, and several remote settlements besides. It was like traveling in the company of a young prince who had set out to seek the will, capabilities, and cultural heritage of his people. At the time, I supposed that Vicha's fervent interest in our travels—almost any excuse would suffice to bring us to some new village or jungle outpost—was a matter of intellectual curiosity, but it is difficult for me to imagine now that he did not consider himself destined for some leadership role in Thailand. Our time together was a sort of pilgrimage for him, in which he sought to find the élan of the people, the source and inspiration of some future role for himself.

Meanwhile our task was to dig up field information so the dam could be designed. The design engineers, who were located in the Denver headquarters of the consulting firm that employed me, had very little information on site conditions at Lam Pra Plerng and were pressing us to provide them with data on subsurface soil and rock characteristics. Soon after my arrival the monsoon rains ended, access and digging conditions became much improved, and Vicha and I pressed forward with the explorations.

Pira, out project manager, encouraged me to use as many laborers as possible in this work, and although I recognized that use of hand labor had a desirable effect on the local economy, my engineering training had burdened me with a parsimonious sense of economy, which always made it difficult for me to assign a crew of thirty men to spend two weeks hewing out a mile-long path through the jungle, all in order to save myself a few minutes of walking time. However, one task that had to be done early in the project was the excavation of a deep exploratory trench in which we hoped to lay bare the underlying rock and soil all along the dam axis, down one abutment, across the flat valley floor, then up the other abutment. It was my idea that we should wait until we had some suitable excavating equipment on the job before undertaking this major task. However, one day the Irrigation Department's director general made a visit to the project and with an impatient sweep of the hand decreed that the work should begin at once.

The next morning and every morning thereafter, I found a crew of a hundred or so laborers picking away at the trench with the hoes that are traditionally used for digging in Thailand, loading the earth into baskets and carrying it off by hand. Some of the excavation was in hard rock, which the laborers broke up. Each laborer, who was paid fifty cents to remove a specifically designated cubic meter of material each day, would light a small charcoal fire on the square meter that he or she was scheduled to excavate the next day, covering the fire with packed earth to transmit the heat to the rock better. The next morning he would pour cold water on the heated rock, which would either crack or weaken it; then for the rest of the day, he would break up the rock with a hammer and carry it away in baskets. About two-thirds of the laborers assigned to this and other heavy tasks were women, and although I was at first appalled by the prison-like conditions of this work, I later came to enjoy my daily visits to the trench, for I have seen few cocktail parties that gave off as convincing an aura of happy, bawdy sociability as a crew of Thai laborers working in 100 degree heat at the bottom of a dusty trench. I have since wondered whether one might not find as much contentment spending a two-week vacation cracking rocks in Yuma, Arizona, rather than abusing one's digestive system on a Caribbean cruise. Perhaps it is really the social conditions involved in our activities that count.

During the course of this work, I noted to my horror that this gaping trench was being excavated several hundred meters away from the true location of the dam and on inquiry was told that the location of the trench had been selected on the basis of everyone's best estimate of where the director general had been pointing at the moment of his decree. The director general was an important and wise man. It was only with the greatest difficulty that I persuaded the engineer in charge to move the trench to its correct location, beneath the dam.

In addition to exploring the foundation, part of our assignment was to locate and prove out the existence of sufficient quantities of earth to build the dam embankment. Of the several areas that I inspected, the most promising source appeared to be a flat terrace flanking the river one-half mile downstream of the dam site. This was also the site of the principal village in the area, Bu Hua Chan. I felt uncomfortable with the idea of wiping out this idyllic settlement of about fifty families for the sake of the dam. However, I planned an extensive grid of exploratory auger holes for the area, and when the results were in, it appeared that the village and surrounding areas were the only practical source of adequate earth materials for the job.

Pira assigned Vicha to negotiate the purchase of the land

homes from the villagers, and we went to great lengths to ensure that the procedure was handled with appropriate ceremony and planning; experience on earlier projects had convinced Vicha that simple cash payment to the villagers was socially undesirable. The idea was that the villagers, unaccustomed to such sudden wealth, would be inclined to spend it quickly and foolishly. Vicha told me of one case where the government had negotiated the purchase of a small, remotely located village within the area that would be flooded by another reservoir. The villagers had spent the money they were paid for their homes on prestigious trinkets such as expensive wrist watches and transistor radios but made no effort to move from the area. Government authorities soon realized that the villagers had simply not believed their story about the reservoir's flooding out their homes. After all their fathers and grandfathers had remained high and dry at this same site, and it seemed to them improbable that the government engineers would be able to change the course of nature so drastically.

Vicha and I visited the condemned village of Bu Hua Chan frequently, he to conduct negotiations with the headman and I to inspect progress on my program of exploratory augering of test holes. My objective in this program was to determine the depth and type of soil that existed in the area. For this purpose, I had a crew of three laborers, and we equipped them with a six-inch diameter hand soil auger of the type that can be purchased in most American hardware stores. Using a compass and tape, I would stake out in advance a grid of hole locations, the holes spaced fifty meters apart. After a while, I found that the leader of the crew had caught on to my system of laying out holes, and I turned this task over to him. This procedure for soil exploration proved simple and foolproof in practice, and I have used it often on other projects where only hand labor was available. It occurred to me at the time that given a large enough labor force and sufficient augers, I could have in this manner established a neat gridwork of exploratory holes covering all of Southeast Asia.

Toward the end of this program, when I plotted the locations of my holes on a map of the area, I found a gap in one place adjacent to the village, and upon inquiring as to the reason for this gap, was told that the crew had not wished to auger there because it was the village burial ground. It was not that they objected to holes being drilled in this area but rather that they feared going there because of the presence of spirits. Vicha, who was well educated, had nonetheless provided me with such articulate and convincing arguments favoring both the existence and activity of spirits that I had come to half-believe in them, and I found visiting the burial area, a dank and gloomy grove of bamboo, a distinctly uncomfortable experience. I decided that perhaps it would be best to leave this area out of our excavation plans, but when I discussed this with Pira, he determined that this would be impractical and arranged for the exorcisation of the place by a delegation of locally prominent monks. This accomplished, we completed our exploration of the area, which proved to contain suitable soils, and it also was later used as a source of embankment material. In this way, the embankment of the Lam Pra Plerng Dam came to contain the bones of several generations of Bu Hua Chan villagers, an admixture that, presuming the efficacy of the process of spiritual disarmament by the monks, should not adversely affect its performance.

During the time we worked together, Vicha seemed engaged in a kind of personal quest, a rite of passage. This seemed to take him further away from the upper-class circles in which he had been raised, toward the smaller towns and villages, and finally, during the last few weeks that he was assigned to our project, to a remote settlement in the upper reaches of the Lam Pra Plerng basin. He had been morose and depressed for several weeks before his visit. He was suffering from some sort of asthmatic problem and seemed to believe that he might have cancer. The symptoms seemed psychosomatic to me and correlated well with a growing antagonism between Pira and Vicha. Avoidance of outright hostility is part of the Thai character, and several Thai I knew suffered from various physical symptoms whenever they became involved in disputes. Whatever the cause, a doctor had prescribed a regular dosage of some evil-looking dark fluid that smelled like creosote, and Vicha carried a pint glass bottle of this medicine everywhere he went. I noticed that his gloom lifted temporarily when a hunter told us one day that there were a few *gar*—evil-tempered wild cousin to the placid water buffalo—abroad in the upper Lam Pra Plerng watershed. He talked to me about the *gar* and seemed fascinated at the prospect of hunting them, as if in such an encounter he would discover some fundamental source of energy with which he might change his life.

One day when we were out scouting for deposits of sand and gravel, which we needed to make concrete at the project headworks, one of the villagers told us that an elephant had been killed by an ivory poacher in the upper watershed. The villagers from a nearby settlement had stripped the meat from one side of the elephant but had not been able to turn the carcass over, and the rotting remains had attracted some tigers, which were feeding on it at night. Vicha became oddly excited at this news and suggested we visit the site at once.

The next day, Luay, Vicha, and I borrowed an ancient Springfield rifle from one of the guards at the headworks and drove south on the rutted clay road that wandered back into the mountains, arriving at nightfall in a small village of half a dozen houses, located in a clearing at the edge of a thick forest. There was no doubt that we were at the right place. Wires had been strung up between the houses, and suspended from the wires to dry in the sun were thousands of strips of leathery brown elephant jerky. We introduced ourselves to the headman and were invited to set up our headquarters at the village.

That evening, Vicha, Luay, three or four villagers, and I sat by a smoky fire next to the headman's house and ate a supper of elephant meat and glutinous rice, washed down by rice wine, a sour, fermented, milky brew that was the principal alcoholic beverage in the Thai countryside. Ordinarily, and particularly in such a remote setting, I was acutely conscious of being six feet tall, white, and, as the Thai never failed to note, earning a salary that was a hundred dollars a month more than the official salary of their prime minister. But camp fires are great levelers. There I was conscious of a blurring of these distinctions. The talk, what I could make of it, for no one felt obliged to translate the northeast dialect for me, was mainly concerned with hunting. On such occasions, Luay, who considered himself more capable than Vicha or myself on such matters, assumed an uncharacteristic assertiveness and told stories of his own adventures in a way that suggested that they were as meritorious as Vicha's or my own. This sometimes annoyed Vicha. "I am going to have to talk to Luay," he would tell me when Luay acted above his place. But that evening, there seemed to be a general agreement that we were all equals—the young prince, his foreign guest, the servant, and the peasant woodsmen. Even among the Thai, I came to realize, class distinctions were recognized as being defined by social convention and were apt to fall away at night in the deep forest, proof to me that there were no labels firmly affixed to the human soul.

Our plan was to spend the next day getting familiar with our camp and surroundings and then, perhaps the day after, to strike out into the forest to find the dead elephant and whatever else might be there. So the next morning we slept late, until 8 o'clock or so, when the damp, cool air that blanketed the forest in the early morning had begun to burn off. Then we arose for a breakfast of tea and a portion of rancid geng, the meat of the small barking deer that was abundant in the forest and considered a delicacy by the Thai. In accordance with the dictates of his mysterious program, Vicha decided to spend the morning in camp learning something about rural Thai cuisine.

I had with me a light bolt-action 0.22 caliber rifle of Czech make, a favorite of mine. It was a fine-handling gun, which delivered extreme accuracy exceeding any promise made by its cheap appearance. I suggested that if there were small game about, one or two of the villagers might like to accompany me on a short hunting expedition. Two of the young men who had been at the camp fire the previous evening readily agreed to this proposal, and we struck out down a nearby forested ravine. I had thought that they might wish to try the .22, but they declined my offer and instead brought the homemade muzzle-loading weapons that most Thai villagers in that part of the country owned. These consisted of a three-foot length of standard steel pipe that had been thinned and tapered slightly on a lathe and welded closed at the breech end, then wired to a homemade wooden stock. A small hole, about an eighth of an inch in diameter, was drilled in the side of the pipe at the beech end and a crude spring steel striking hammer fashioned, which would strike the hole when snapped back with the thumb, or, in the more sophisticated models, when released by a trigger. The hunter carried a small cloth or woven twig bottle of black powder, some cloth wadding, and lead balls; the weapon was loaded in the familiar sequence of rammed powder, wadding, and lead ball. A conventional red-cap-pistol paper cap was placed against the hole, and the weapon fired by snapping the hammer against the cap, which sent sparks through the hole, igniting the main powder charge. This exploded with a clap of thunder, a belch of flame, and a puff of white smoke. This crude musket was not very accurate, and its effective use required the hunter to stalk within a very few yards of his target. The elephant reportedly had been killed with one of these weapons, and I had to admire the courage of the man who did it. Reloading the weapon was a time-consuming business, and one would have no second chance if the elephant were merely wounded and decided to seek revenge.

There were many squirrels in this part of the forest, and for most of the rest of the morning, the three of us walked within cool glades, stopping from time to time in an attempt to lure the squirrels down from their hiding places in the jungle canopy seventy or eight feet above our heads. The villagers had a novel technique for attracting the squirrels, which I had not seen before and which I have subsequently found is ineffective on our North American squirrels. They would grasp a low-hanging tree branch and beat it against the ground, simultaneously thumping the ground with their feet. For some reason, any nearby squirrels would become much annoyed by this sound and would respond with an angry chatter, and after a few minutes, they would approach the source of the noise. In this way, once or twice we were able to lure a squirrel out of the foliage and onto one of the gray, smooth trunks of the high trees, but they were always careful to stay on the other side of the tree. Once by quietly circling a tree, I was able to get off two shots at one of them while it was on the run, but these only splintered the wood a foot or two behind him. This being the best we were able to do in the course of two hours, the villagers said that it was getting too late in the morning, and we walked back to the village. By this time, the forest had grown silent, and now with the heat beginning to seep down through the canopy above, it was a time for sleeping, for the hunters and the hunted.

During the morning I had begun to run a fever, and in

the absence of the usual stabbing cramps that heralded the onset of dysentery, to which I was well accustomed, I was afraid that I might be coming down with an attack of malaria because the day before I had neglected to take my usual regular dose of suppressant medication. Luay was a regular malaria sufferer. For reasons of his own, he preferred a routine three-day bout, with its mechanically regular sequence of intoxicating fever, chills, and sweats, to taking preventive medication. I was therefore constantly in the company of a rich reservoir of *Plasmodium vivax*, the protozoa that causes the disease, and although I kept a sharp eye out for its carrier, the anopheles mosquito, with its peculiar poised biting posture, I could never be sure that one of the daily dozen or so bites that I got off with on my best days was not the one that would trigger an attack.

My recollection of the rest of that day is impressionistic and sensuous—a leaden sleep on the hard bamboo mat at the headman's house, waking in the heat of the afternoon, a curious and somehow comic difficulty in opening mucilaginous eyes and mouth; a decision made by someone, perhaps myself, that I should go to the hospital in Korat; and the long ride in my squeaky, pitching, green jeep, Luay driving, and me vomiting in the cool evening. It was a curiously pleasant afternoon, heady and dreamlike, in which, I felt a total absence of responsibility and a sense of comfort and security in the total possession of my body and brain by the all-powerful fever demon as the controls slipped from my hands. I have since taken consolation in the thought that if death brings with it that exhilarating, orgasmic loss of care, it cannot be an unpleasant experience.

In this way, I descended into a sweat-soaked purgatory of headaches and clockwork fever cycles that lasted for seven days, at the end of which I arose to resume my daily routine at the project with a sense of redemption and rebirth. Vicha, it turned out, had attained his own mystical experience, and I had to confess to some envy when I heard of it. The night that I had returned to Korat, he had, in company with my two hunting companions, hiked two hours or so to the site of the puffed, malodorous corpse of the elephant. At nightfall they climbed a tree twenty yards or so away from the carcass, and, tired from their walk, fell uncomfortably asleep on a broad limb fifteen feet above the ground. At about two o'clock in the morning, Vicha awoke suddenly to a deep groan and after a few moments became aware of something moving through the brush a few feet away. A moment later, on the other side of the elephant, there was the unmistakable mucoid gurgle of a big cat. Two tigers had come to feed on the elephant. With the realization that hostile forces were about, the tigers had adopted a strategy of reconnaissance in which one of them remained alert and breathing heavily below the tree and the other circled. This had gone on for at least an hour.

"I was shaking so much that the tree shook, and it was very noisy," Vicha told me, laughing, his eyes wide with a look of vitality I had not seen in weeks. Then in the still blackness just before the first light of dawn, both tigers closed in on the elephant and, with a noisy tearing and cracking of leathery skin and bones, fed for a few minutes. Vicha put the flashlight on the scene, and at once one of them glided silkily off into the jungle, "But the other looked at us with bright eyes," Vicha told me. "I fired at him with the shotgun, but I was not steady, and it was a wild shot." With a great leap, the tiger was gone.

A week or so after this trip, Vicha was transferred permanently to the Irrigation Department headquarters in Bangkok, a transfer I heard he had arranged himself through his father's influence. He had, it seemed, achieved whatever objective he had set for himself, and now he was ready for the more sophisticated politics and intrigue to be found in the capitol. Neither Vicha nor Pira regretted the move. "That Vicha," Pira told me with unconcealed contempt, "he is a very big playboy." Vicha's departure left Pira's project with a cadre of project officers uncontaminated by anyone with an independent base of power and influence, and Pira seemed pleased. As for me, only a couple of months into a two-year assignment, I felt a growing emptiness. What had been appropriate props in the drama of Vicha's quest—and vicariously, my own—the sticky heat, the slow progress in our work, the slanting sheets of monsoon rain, now became irritations. Vicha had declared his independence, thrown off Pira's harsh domination. He had pushed out of a sort of harbor that we had shared. Now beyond the narrows, he was at sea, with a full view of new coastlines. I had miscalculated in my approach, turned back, and was envious. It was to be a while, several years, before my chance to run the passage would come again.

If Pira's green jeep provided me with an immediate if undeserved status on the project, Luay, who came with the jeep, proved an even more lasting legacy. Luay had been Pira's personal driver for several years. It was a position that had carried no little prestige—and prestige meant something to Luay—so he at first rebelled against the transfer, and it became Vicha's job to deal with Luay's protesting sulks and absenteeism during his first few weeks in my employment. For my part, I spurned the idea of a chauffeur; I could drive myself, I told Vicha, and moreover the whole idea of a personal driver seemed repugnantly undemocratic to me. I was an engineer, not a pharaoh. "Yes, yes, you can drive yourself, but Luay can ride in the back; if you need some thing, to get some cigarettes, or to send that message, he can do it for you," Vicha would say to me, urging that I accept the arrangement. "Thai custom," was his final word, as usual. So after a while I gave in, and Luay gave in too. I suspect that Pira's mandarin irritability had been difficult to live with, and Luay, a cheerful man, seemed happy in the end with the change. He remained with me the whole time I was in Thailand and he has remained much in my mind since.

Luay was a short, compact man about forty years of age. He had a bright look about him, a fine smile, and clear brown skin stretched tightly over a knotty muscular frame. He had a certain bearing, a dignity in his movements and gestures, qualities of which we in the West are barely conscious but which are as obvious to the Thai as the essential tonal dimension of their language. Luay could quickly repair the most vicious blowout or a herniated fuel pump in the middle of a muddy road when the temperature and humidity were both ninety-two, and afterward his navy blue shirt and pants would still be freshly pressed and his black hair would be as neat as if he had just come from a barber. If I now close my eyes and imagine Luay, I see him walking, perfectly straight, slightly forward on his toes, then suddenly freezing, alert, then turning to me, raising a finger and smiling, naming the bird or animal that had cried out in the forest, then imitating its cry and telling me its Thai name and its habits.

Luay was intelligent and loyal. He had what people used to call character. He was not especially impressed with my whiteness, affluence, or college degree, so there was no reason for him to become later disillusioned with those illusory assets. He remembered diving into a river to escape being strafed by a Japanese airplane, and he was married and had two children. On these bases I granted him more experience with life than I had. Over the months I dropped my usual precautionary vigilance so that by the end of a year, Luay no doubt knew me better than I knew myself. Because his opinions and judgments had arisen in such a different cultural context than my own, they carried and still carry a special weight for me, much more than the views of authorities who shared my background and who were therefore subject to the same fads and had the same blind spots as I. For example, Luay hated to throw things away, and he did not like me to throw things away either. This was not just a matter of his relative poverty, as I first thought; it was not that Luay was desperate enough to take scraps. Rather he thought that throwing something away showed a lack of imagination. A clever person would put old boxes, packing materials, and pieces of string to beneficial use. Only stupid people would waste things. The profound significance of this simple truth has become apparent to all of us only in the past few years, but Luay knew it all along. In this and other matters, Luay's outlook served as a sort of practical and moral benchmark, and although I might not agree entirely with his opinions on money, wives and sons, Cambodians, death, and virility, I recognized that they were founded in an alternate culture and that I could benefit by observing the differences in our outlooks. For just as our surveyors arbitrarily accept as fixed a benchmark that they well know is whirling through space at prodigious speeds, so in the moral universe it is only differences that matter most.

Although I came to know other village Thai, like Longmah, who had less exposure to Western ways than Luay had and who therefore might have been better qualified to fill the role of noble savage, Luay's exposure to jeeps and polyester did not seem to have seriously contaminated his Thai ways, and I believe that he was spiritually and culturally Eastern in his habits, typical of a citizen of the non-Western world in 1963. True, he did not conform to that exotically Oriental image held by certain piously romantic European and American intellectuals. As far as I was able to tell he did not speak in nonlinear riddles. But then neither did I belong to the American Nazi party nor drive a Cadillac at Miami Beach nor chew bubble gum. Luay and I were both what you might call middle-of-the-road citizens of the world, and it was not our differences that were remarkable to me but rather that we had so much in common.

Luay moved his family—his wife, Yoi, and children, Mahlee and Vinai—into my house for a few months. During their stay I derived some impression of Thai domestic life, from which I conclude that it did not differ in any fundamental respect from family life in Palo Alto, California. True, there were certain differences in style. For example, Luay and Yoi slept, out of preference, on bamboo mats rather than waterbeds. As with the marriages of my friends, it was never entirely clear to me who was in charge. Luay had his bad days and his good days, which when superimposed on the independent cycles of Yoi's bad and good days, produced the usual marital seismogram unique to them but statistically similar to all other marriages. Neither Yoi nor Luay was much given to formal religion; I do not believe that Luay had ever done service as a monk, just as I know many men who have never attended a weekend at Esalen.

Yoi was an attractive, intelligent, ambitious woman who was frustrated because she spent most of her time at home taking care of the children. She would have been happier with a job or, better, running her own business. As a result she became depressed and suffered from headaches and various ailments. Yoi blamed this on the fact that someone had put the evil eye on her, and she sought treatment from a quack doctor who prescribed tranquilizers and opiates. At these times Luay went about muttering the Thai equivalent of "my wife is impossible." If Yoi had been a forty-year-old housewife with a degree from the University of California and two children underfoot, her ills would have been no different, although she might be attributing her symptoms to frustrations arising from some appropriately psychiatric origin instead of the evil eye.

Luay took refuge from domestic strain by going off on a monthly hunting trip, from which he usually returned trembling with malarial fever. Yoi accepted but did not like this practice. Luay's monthly hunting trip provided him with a great store of conversation, for he loved to talk of stalking and of the habits of animals. As with the best of hunters, the technology of the sport, the guns and ammunition, was a small part of the business. The real pleasure was in the drama of life and death held on a dark, silent stage of a deep forest. The final scene, featuring shining desperate eyes, sudden lunges, animal screams, a confusion of darting flashlight beams, and finally a shot, was a minor part of the play.

In those days I was a good marksman. Why this was so has always been a great mystery to me, for I possess none of the qualities—keen eyesight, a steady hand, calm temperament, or great powers of concentration—that one might associate with such a skill. Nonetheless I was once captain of the 588th Engineer Battalion pistol team, and I could stand on the porch of the bachelor engineers' quarters at Lam Pra Plerng and consistently shoot beer bottles that the Thai could not hit. Luay, who preferred to hunt with a shotgun, made a great pretense of admiring my skill. He would say to me, "Nai shoots rifle, oh so very accurately, can shoot barking deer from very far; I must get very close, shoot WAANNGG! with shotgun. Nai very good shot." This was Luay's way of engaging in nominal flattery while at the same time pointing out my incompetence at creeping and stalking, the real stuff of hunting.

One day soon after I came out to Lam Pra Plerng, Vicha, Luay, and I walked down to Bu Hua Chan to look at some soil samples in a test pit that I had arranged to be excavated next to the little temple at the edge of the village. There we happened to see the assistant village headman. We stopped to talk with Longmah about our soil exploration program, and the conversation soon turned to hunting. Longmah said that there were tigers and elephants in the hills, back above the dam, but that it was illegal to take

128   them. Nearer the village one could find monkey and rac-
coons—not very good to eat—but also wild pigs. The men
in the village liked to hunt pigs, Longmah said, and he
went on to describe how one of them had been gored in
the groin by a wounded boar. Longmah was small, thir-
tyish, and wore a pair of ragged khaki pants and a pacoma,
the towel-size checked cloth that serves as hat, blanket,
bathing suit, rope, towel, umbrella, and belt in rural Thailand. He paused and sized me up for a moment. Would the
foreigner like to go hunting for pigs? Vicha, Luay, and I
agreed that he would, and we arranged to meet early the
next morning. That evening Vicha and I borrowed a couple
of shotguns from the project guards and at six o'clock the
next morning, the four of us met at the edge of Bu Hua
Chan. For an hour or so we hiked on a trail through pad-
dylands and then into the forest. After a while we came to
a patch of thick bamboo, where we entered to take up am-
bush positions as directed by Longmah. I remember that
Luay and Longmah were able to slip through the bamboo
at walking speed. As I crashed along behind them, I heard
Luay complain to Longmah about all the noise I made. "It
doesn't matter; they'll think it's an elephant," said Longmah. Longmah led me into a dense stand of bamboo and
told me to wait. There I remained squatting and sweating
in that miserable patch of sharp sticks for an hour or so
while Longmah and two other villagers, who had mysteriously appeared, shouted and beat around the edge of the
bamboo grove hoping to drive some pigs in our direction.
I could not see more than a yard ahead of me, and though
I once heard something crashing through the brush only a
few feet away, I had no idea what it was. Of much more
immediate concern were the carniverous red ants that
kept crawling up my pants. I was glad when the hunt was
terminated and did not trouble to follow Luay and Longmah's detailed analysis of the event.

Of greater interest to me was the visit we then made to

129 the nearby house of Longmah's uncle, which stood beneath a magnificent banyon tree at the edge of the forest. Thai village houses are raised on stilts about eight feet off the ground and as customary for visitors, we removed our shoes, climbed the ladder, and sat cross-legged on bamboo mats on the hardwood floor of the platform. The uncle was the official host, and his wife remained somewhat in the background, puttering about with various utensils, but listening carefully and commenting from time to time. They seemed pleased to have such unusual guests, and we exercised our mutual curiosities. How old was I, he asked. How old did he think I was? The uncle inspected me carefully. Sixty, he said. I told him I was only twenty-four and asked him how he had decided I was sixty. "Because you are so big," he said. We talked about the weather. The Thai like to talk about the weather even more than we do and will say by way of greeting, "Looks like rain," even though it is the middle of the dry season and everyone knows that it will not rain a drop for two months. I asked the uncle whether living so far from the village was safe. He replied that usually there was no problem but that two years previously they had been raided by bandits. I asked Vicha whether he thought the bandits were actually Communist guerrillas, for according to the U.S. press, northeast Thailand was hopping with Communists. He did not think so; more likely they were army deserters.

The uncle asked us if we would like to eat, gin kow, which is the colloquial Thai for "eat." Literally it means "take rice." Oh yes we would indeed, Vicha said, telling me in English that I should accept the invitation so I would learn about Thai ways. Longmah charged up his muzzle loader and shot the head off one of the chickens that was squawking around under the house, and we promised to come back in an hour for lunch, after scouting around the area for deposits of laterite gravel for road building. Later that afternoon as the four of us walked back to Bu Hua

Chan, Vicha encouraged me to converse with Longmah, coaching me in my crude Thai and translating when matters got confused. Longmah always spoke in a very loud voice to me, in the manner of an American tourist visiting Europe for the first time in 1950. I cannot recall the details of our conversation that hot afternoon so many years ago, but I remember how pleased I was to be conversing socially with someone who lived on the opposite side of the globe. Even more pleasing and flattering to me was Longmah's appearance at my soils laboratory the next day, when he informed Vicha that he would like to work for me. Vicha subsequently made arrangements to have Longmah hired as a laborer-technician. I taught him to run field density tests and to use a slide rule, and he was of great help to us as a guide and ambassador to various villagers who lived in the area affected by our construction operations.

Luay did not like it when Vicha and I paid too much attention to Longmah, and when Longmah was not present Luay belittled him. Longmah was a hick. Luay snickered at Longmah's provincial speech—Longmah spoke Lao, he told me once—at his primitive muzzle loader, and at his peasant clothing, his lack of a vehicle. It was the same attitude that we summer kids from Boston used to hold toward the Cape Cod locals.

I liked Longmah and tried to open up opportunities for him on the project. I do not think Luay much approved of this interference in local affairs, and perhaps in a way he was right. In the end he extracted a kind of minor revenge on Longmah for his arrogance. It was a small but significant incident that took place my last afternoon at Lam Pra Plerng. I had said good-bye to the technicians at my soils laboratory, and Luay and I were driving on the dirt road toward the main gate of the camp. Luay was driving the jeep, and we came up on a Thai peasant who was walking on the edge of the road. Luay was driving fast, and we roared past the man, churning up a great cloud of dust. It was true that Thai drivers were not finicky about near misses of roadside pedestrians, for the custom was that the road belonged to machines, not people. But in this case it seemed to me that Luay cut unnecessarily close with rather more speed than necessary. I looked back, just in time to see the man remove his *pacoma* from his face to see what had almost run him down. It was Longmah. I opened my mouth to speak, to tell Luay to stop, so I could say good-bye to Longmah, perhaps to give him something like my American jackknife, which he had admired, just as I had admired his homemade one. But Luay started to laugh and said "Longmah" in his most derisive manner, and I did not say anything. I was afraid I would not know what to say to Longmah, afraid of what Luay would think of such a frankly sentimental gesture. It was not Thai custom to do such things. So we sped on, and my last sight of Longmah was of a lone peasant on the road disappearing into our brown plume of dust. It was a poor farewell.

In the early 1960s the U.S. State Department persuaded a prominent Thai religious leader to take a poll of several dozen of his countrymen. The survey asked the Thai to identify any behavior typical of Americans that the Thai found offensive and to provide, to Americans unfamiliar with Thai ways, advice on avoiding cultural blunders. With remarkable consistency, most respondents to the survey provided the following three pieces of advice:

1. Don't put feet up on desks or point them at people.
2. Images of the Lord Buddha should not be used as house decorations.
3. It is not Thai custom to fondle women in public.

There is more to the last point than stated. The Thai may be congenial, fun loving, social, and even emotionally expressive, but they do not like to be touched. Backslapping, arm grabbing, and (especially) hair tousling are grossly offensive to the Thai, for whom the head is sacred, the feet profane. Touching is generally felt to be an invasion of privacy in a crowded society where there is little privacy from the eye. The traditional Thai greeting is the graceful *wai*, in which the head is bowed while the hands are held in the Western praying fashion. Sophisticated Thai in Bangkok will shake hands with foreigners, but the shake is perfunctory and limp, the hand quickly withdrawn. Pushing or shoving, any physical violence directed at the person, will stop traffic in Bangkok. Given the formality of the master-servant relationship, physical contact in that context is all the more unusual. For that reason the following two incidents involving Luay stand out clearly in my mind.

The first incident occurred in the dry season on the edge of the dusty road between the silk-weaving village of Pak Tong Chai and the town of Korat. Luay and I were standing beside the jeep. My right hand, in which I was holding an automatic pistol, was resting on the canvas roof of the jeep. I was aiming at a crow that had just landed on the branch of a tree about 250 feet away. Several other crows were circling around the tree. It was late in the afternoon, and the dry, stubbly rice paddies rippled in the heat. An ugly metallic smell of gunfire hung in the still air. Moments before I had shot and killed a crow perched on the top of the tree. It was the very top of the tree. The shot was incredible. I had aimed about two feet above the crow and then squeezed the trigger. The pistol jerked. There had been a long moment in which time itself seemed to have swooned in the heat. Then as the rolling echo of the shot had returned from the plain, the crow had fallen gracelessly from the tree, and Luay had cried out, "Oh Nai, dee mahk," which means, "Oh, boss, very good." The Colt had a magazine containing eight shots. There were seven left, and I was about to repeat the feat. Then perhaps I would do it again, and again. . . . A hand touched my elbow. It was Luay. He was shaking his head. "The crow is a poor bird," he said.

The other moment occurred several months later in a new shopping center that catered to Bangkok's rapidly growing American community, AID people, military advisers, businessmen and their families. We were double parked in front of a new pizza restaurant. Two American high school boys, teeth in braces, came out of the restaurant and climbed onto a motorcycle. Luay opened the canvas flap on the back of the jeep, and I removed my dusty Samsonite suitcase. Only this morning we had left Lam Pra Plerng, drove out of the hills, past the remains of Bu Hua Chan, the paddy fields now green and wet, to Korat, then down the Friendship Highway to Bangkok, on the steaming Chao Phraya plain. Tomorrow I would leave Thailand for good, but in some ways I felt home already. Luay, who suddenly looked smaller and browner to me, seemed out of place in front of the pizza restaurant. The day before I had given him my shotgun shells and a wool shirt, and I was not sure what to say now, so I stood there foolishly on the sidewalk, facing this man with whom I had spent most of the days of the past two years. Luay grinned and stepped up to me, reached out and shook my hand.

When I first arrived at Lam Pra Plerng, several kilometers of the main irrigation canal had been excavated, and work had begun on some of the minor concrete siphons, flumes, and drop structures that were required wherever the canal traversed any irregular terrain on its course across the paddy fields. The main canal itself was a five-foot-deep ditch dug into the iron-red and black clay soils of the river floodplain. This excavation was accomplished by an ancient, track-mounted dragline, which advanced the canal a few tens of feet each day. Creaking and squealing, this machine dug into the soil bucketful by bucketful, heaping and patting the earth excavated from the ditch into banks on both sides of it. These banks had a design purpose; when necessary, the canal could be temporarily overfilled with water without flooding the adjacent farmlands. Conversely the banks would keep any floodwaters from the nearby river from pouring into the canal and filling it with mud.

Now conventional construction practice would require that watertight earth banks be carefully built up of layers of earth, each layer being thoroughly tamped by heavy rollers, with the result that the banks would become stable dikes capable of resisting the erosion of swirling floodwaters. But in those early days, we had only the dragline on the project, and although its operator was an artist who produced very fine earth banks by skillfully manipulating his ten tons of rusty machinery, I observed that a few inches below their smooth surface, the canal banks were no more than loosely heaped clods of earth that would wash away in the first heavy rain. In order to demonstrate this flaw to all concerned, I performed several standard field density tests, each of which involved excavating a small hole in the bank about the size of half a grapefruit and then refilling the hole with a standard sand of known density. By comparing the weight of the standard sand that goes into the hole with the weight of the clay that comes out of the hole, one can evaluate whether the clay is suitably compacted or excessively loose and porous. This test, which is performed worldwide as a means of checking on contractors' earthwork construction quality, has a rational and scientific basis. However, the real secret of its effectiveness is unwritten in any construction inspection manual. The giveaway is the special sand that is used, a silvery white sand mined from a glacial beach deposit near Ottawa, Canada. In fact, there is no good practical reason why some yellow or brown local sand could not be used in the test. But this would rob the ritual of its magic, and the results would not be nearly as awesomely authoritative. All of this first became apparent to me only as I performed the test at various locations along the canal banks at Lam Pra Plerng. Local villagers would gather around me, watching my priestly manipulations with the glazed look of people about to receive holy communion. A foreigner has come into their midst, paid, it is said, even more than the prime minister. He wanders about the project works, kneeling to dig little holes in the ground, then refilling them from a glass ciborium with an icy white sand mined on the other side of the earth. He manipulates runes on a white bamboo stick and then renders a judgment as to the goodness of the earth. Few Thai villagers or American contractors would dare to doubt the authority of a verdict accompanied by such solemn rites. So we embellish our professional activities with magic, disguised in one way or another as rational.

Most of my tests on the canal embankment indicated substandard compaction, and when I presented the results to Pira, our project manager, he wisely suggested that I could do a far greater service to the project by devising a means of improving the quality of the banks than by forwarding the test results to Bangkok, where they would cause some major bureaucratic breakdown. I readily agreed to this constructive approach and set about devising a backyard method for compacting the embankments. At first I considered the possibility of using hand labor, which was in plentiful supply and was a socially desirable solution to any problem because it created jobs. But delivery of the enormous energy required to compact the tough clay soil adequately by hand appeared to be a hopeless task. I calculated that a human energy output equivalent to what was required for a person to climb Mount Everest would yield compacted soil sufficient for only about a three-foot length of canal.

It happened that at that time the Irrigation Department was performing an operation described in official reports as "clearing the reservoir" but which in fact consisted of the more profitable enterprise of cutting and removing the few select hardwood trees that existed in the area due for eventual flooding by the lake. In accordance with usual practice in Thailand, specially trained elephants were used to drag this commercially valuable timber from the forest. Elephants grow tired of working after a few days and need vacations. Our project elephants took their periods of rest and recreation in a grassy meadow directly behind my bungalow. At the time we were attempting to solve the canal embankment problem, two bulls, a cow, and a month-old calf were in residence in this meadow. I have always been fond of elephants, and I remember how it gave me much pleasure to sit behind my house in the early evening and watch these animals contentedly putter about. It is difficult not to feel kinship with elephants, and I sometimes passed time by watching their features for expressions of amusement, fond recollections, or troubled consciences.

I had a small vegetable patch behind my quarters where I raised lettuce and corn, and I reckoned that the elephants, regardless of their conscience, would make short work of this if they discovered it. One of the boy elephant keepers had told me that a single strand of telephone wire strung horizontally a meter above the ground would be sufficient to keep the elephants out of the garden. This was just the kind of admirably clever solution one could expect from elephant boys, who are able to establish, solely on the basis of clever bluffing, a benign but absolute tyranny over their charges. It is a comic sight to see one of these five-ton beasts caught in some act of misbehavior, cowering under the blows and shouts of an eighty-pound boy. The trick of confining the animals with a single strand of wire, which I suppose was effective because it took advantage of a cluster of visual and psychological weaknesses that elephants have, seemed an appropriate solution to the problem of protecting my garden.

I was in the process of installing such a barrier one evening when it occurred to me that the elephants might provide the perfect solution for compaction of the soils in the canal embankment. Certainly they had the necessary weight. The idea attracted me because I thought that it might prove to be a solution that was elegant as well as practical, and I knew that it would please the Thai, who are fond of elephants and would be delighted if they were better at the job than a spread of clanking imported machinery. Accordingly I had no difficulty in persuading Vicha to help me test the idea by setting up a field trial of the elephants' compactive ability. We arranged to obtain a few truckloads of the canal embankment soil and for public relations purposes set up a simulated embankment next to the project headquarters. We placed the soil in its usual loose, uncompacted state and performed the sand-cone density tests on it, then tramped the elephants back and forth across the test section a few times, retested the soil, then repeated the process a few more times, the idea being to show the progressive improvement of the compacted soil.

Unfortunately the results of the test demonstrated conclusively that elephants were entirely ineffective in compacting soil, for two reasons. First, in spite of their weight, the animals almost always distribute it more or less equally to three feet, with the result that the unit pressure delivered to the soil is quite low—on the order of 54 psi. Second, elephants soon come to understand that compacting soil is work and they quickly learn to walk in their previous footsteps. The end result is a narrow, beaten elephant track rather than a uniform compactive effort. Strenuous bullying by the handlers is not sufficient to produce uniform coverage of the test section.

Disappointed with the results of our test, we abandoned the plan of using elephants as compactors and in fact never did develop a satisfactory solution to the canal embankment problem. However, it occurred to me that the results of our test program might provide a curious if trivial footnote in the archives of technology, and so I wrote a short paper on our research. One of the Thai engineers in my lab translated it into Thai, and it subsequently was published in the journal of the Thai National Engineering Society. Vicha told me that in accordance with the custom of the society, we would be asked to make a presentation of our findings at one of its monthly meetings a month or two later. However, I never was invited to any meetings of the society, and when I made inquiries regarding the reason for this, I was informed by the Thai, with a great deal of embarrassment, that they thought I might be offended by the obscene show that was ordinarily presented during the last half of the meeting.

Recently, and much to my amusement, the results of this odd experiment surfaced again in the New Yorker magazine. This came about in the following way. A few years after I returned to the States, I rewrote my paper in the form of a brief technical note entitled, "The Uselessness of Elephants in Compacting Earth Fill." The note was written intentionally as a parody of the usual technical note, complete with solemn assurances of experimental controls ("a soils technician was assigned to ride each elephant at all times during the test.") and parenthetical footnotes regarding the historical use of elephants in oriental warfare. It was published in the *Canadian Geotechnical Journal*, the editor of which, Vic Milligan, was a man with a sense of humor. However, the note was later republished in abbreviated form by someone who did not get the joke. The New Yorker picked up this abstracted version as an example of technical trivia, noting, "Centipedes probably wouldn't work either," thereby missing the point or at least the comic intention of the original article. For once, the *New Yorker* did not get the joke. But I was pleased anyway that the experiment was still providing, years later and in a fourth reincarnation, some amusing diversion to a mass audience of sophisticates far removed from the original action.

In my early days at Lam Pra Plerng I had been delighted by the cross-cultural rapport that had arisen between my partner Vicha and me; we saw and felt about so many things the same way. When he left Lam Pra Plerng that autumn, I missed that companionship. My work seemed less exciting then, as the dry season settled over the Northeast, and the color began to bleach from the Thai landscape.

But I was busy drilling exploratory holes at the dam site, and I soon became friendly with the technician who prepared geological logs of the bluish-green rock cores that our drilling machine extracted from deep holes. Nirand was his name. He was a young Chinese, son of a merchant from a small town near the Cambodian border. He was studying English, he said, and would like to learn very much by talking with me. On my daily visits to the drill rig, Nirand never failed to attract my attention with some clever speech or announcement. He was a young man on the make, and I recognized that, but I liked him for it. He was smart and ambitious, and he openly admired what we used to call the American way.

I once tried to photograph our drilling operations, the fine old drilling machine and the crew of laborers who scrambled over the rig performing various photogenic activities. Unfortunately I was not able to convey the concept of an action shot to the chief driller, who, whenever he saw me with a camera ordered that all operations halt and organized his crew into an elaborate and hierarchically correct pose, like an old-time sculling crew. I still have that picture, which has the look of a nineteenth-century daguerrotype: the fine gasoline-powered engine, with its polished fly wheel and rubber drive belts, the raggedy peasant drill crew, posed stiffly with various tools, and the scratchy blurred effect invariably achieved when I had my film processed by the photographer in town. Off to the side, separated from the group physically, outside of the chain of command, squats Nirand, a sort of minor league scholar with his clipboard and pencil. The driller and his crew are Thai, and Nirand is Chinese, a member of a racial minority that controls much of Thailand's economy. Nirand, a small town boy, lives and works with the Thai crew in apparent harmony, the same polite and somehow suspect détente that I remembered prevailed between Jews and Protestant Yankees in high school. (We Irish, fellow minorities, were always either close friends or bitter enemies with the Jews but never merely polite.)

Nirand used to drop by my quarters in the evening, and we would sit smoking crude native cigarettes, fat pinches of tobacco rolled in banana leaves, and talk of many things, partially in English and partially in Thai. We both took great pleasure in exploring and comparing the social nuances and implications of our respective languages. "Why do you call your assistant *nong*," I would ask; "doesn't that mean little brother'?" and Nirand would explain how *nong* did not exactly mean little brother but is an appellation subtly suggesting a certain type of senior-junior relationship among the Thai. I would try to explain to Nirand what the English word *no* means precisely, which is quite difficult to do because in Thai there is no such word, the concept itself being superfluous to the Thai point of view. Then Nirand would explain to me which of the five different Thai words for *I* should be used under what circumstance.

My Thai improved rapidly during those weeks, and I soon developed a facility for using the tonal variations that are essential to the meaning of the langauge: how to say the word *kow* in the different tones required to give it its various meanings of "rice," "mountain," "white," and so on. My ability to mimic the musical sound and rhythm of the language soon exceeded my vocabulary and comprehension, and for the rest of the time I was in Thailand, I invariably mystified Thai villagers by opening conversations with great fluency and then failing to understand much of the exchange.

About this time an incident occurred that soured my friendship with Nirand; the recollection of it makes me shift uncomfortably in my chair. One afternoon as I arrived at the site of the drill rig to inspect and log the samples of the rock core that had been extracted from a drill hole on the right abutment of the dam, the drillers told me with great excitement how earlier in the morning, when they had been setting up the machine to begin work, a large king cobra had reared up from the grass, spread its black shining hood, paused with deadly and beautiful aerodynamic precision, then lunged with the terrifying sound of a punctured tire at the chest of one of the workers who was stacking drill rods. The man jumped back, and while the rest of the crew fled in terror, he swung a piece of drill rod at the head of the snake, knocking it senseless. He finished the job by decapitating it with a machete. This triumphant escape from near death was cause for a party, and the drillers invited me to share in their good luck by attending a feast of the snake with them that very evening.

When I arrived at the corrugated iron shack where the drillers slept, I found Nirand and the rest of the crew sitting by a fire merrily feasting on a curried stew of cobra meat and drinking the raw-tasting Mekong rice whiskey. At a dollar a fifth, Mekong was the drink reserved for special occasions by the Thai. I ate some of the white slippery cobra flesh and then succumbed to what I felt was ribald encouragement on the part of my hosts to get drunk. I should have known better; the scene was moving headlong toward a repetition of the unfortunate evening with my platoon in the army two years before. But I didn't know any better and swilled the caramel-colored whiskey from a glass that the drillers kept happily refilling for me until, after about an hour, my brain wheeled and smoked like a pyrotechnic pinwheel. At this moment, Nirand's wife, a demurely graceful Thai girl of about eighteen, floated before me. She was dressed in the simple cotton *pasim* of the Thai countryside, and the flickering light of the fire fell on the smooth, dusky skin of her arms and face. She felt my stare and lowered her eyes. High on her cheek she had a small scar, an imperfection that served as a unifying link between the ideal and the actual, and which ignited a combustible mixture of gentle reverence and reptilian desire within me. I turned to Nirand and spoke in the level, flat tone that drunks invariably believe will sound sincere and reasonable.

"Nirand, I want to sleep with your wife," I said, in English.

Nirand looked at me, puzzled. "I am sorry, sir, I do not understand," he said. Under pressure, Nirand reverted to the formality of an old-fashioned English copy book. The drama of the exchange had been instantly picked up by everyone, and all talk suddenly stopped. The fire hissed with delight. With the reckless exhilaration of the fully committed, I repeated my request, louder.

There was a long silence, then some whispering. The party broke up a few minutes later, and I somehow managed to get back to my bed.

The next morning, I arose late and made my way in the sweltering heat to my office, where I sat ineffectually shuffling some damp papers for about an hour. My brain felt as if it had been half-eaten away by termites, and my pores oozed an oily broth, residue of the previous evening's whiskey.

Nirand showed up at my office as I had expected, late in

the morning, and he fixed on me a shocked and mournful look that was much harder to take than the anger I had expected. He said, "Good morning, sir," and then sat tensely and without speaking in one of the uncomfortable wooden chairs in my office. After a long silence I heard myself speaking, with a slight stutter and a Boston accent.

"I'm sorry about last night; I acted in a stupid way, and I feel very badly about it. Thai whiskey is very strong." I said this last with a pained smile that begged forgiveness.

Nirand looked even more mournful. Well, I thought testily, perhaps he would smile if he did not have a hangover, too.

"I'm glad there was no one there who understands English," I said brightly, hoping that had been the case. Deep in the vault of my conscience I could hear the futile clicking of a switch, long disconnected, that once would have fired volleys of Hail Marys, batteries of rosaries.

"I too wish that had been so, sir," said Nirand, "but I regret that Mr. Chamlong, he understood. Everyone knows." I understood then that Nirand had spent the morning sifting the ashes of his carefully developed reputation. He was down now, and the Thai would be on him like hyenas. That's what you get for consorting with foreigners, they would say. It was not going to do my reputation in camp much good either. I looked at the pistol that lay on the corner of my desk, for the rats that invaded my office to eat the bindings of my soil mechanics journals, and I thought of poor Ferguson, drunk and waving his gun on the train from Kalasin.

In the following months, Nirand and I maintained a polite working relationship, and from time to time I hoped but did not believe that he had dismissed the incident. His English improved steadily, and I took upon myself the role of an exacting schoolmaster. A few months later, the chief of the American detachment of military police at the Korat Air Force base, an army major, told me that he needed an interpreter, and I sent Nirand to him with the highest possible references. A few days later Nirand quit the Royal Thai Irrigation Department and went to work for the U.S. Army with a large raise in pay. I saw little of him after that, but I suspect that he has done well. Sometimes I imagine that he recalls the night we feasted on the cobra as a fortuitous accident that advanced his career. Perhaps. Ambition comes first with the Chinese, the Thai always told me.

In Thailand the end of the monsoon is followed by the dry season, which becomes, over a time of months and at the inexorable rate of a tenth of a degree centigrade each day, the hot season. It is a time of withering, without shadows, on the Korat plateau, when comfort comes only after sundown.

But the passage from day to night comes with quick mercy in the lower latitudes. In minutes, the heat and light, which for the last twelve hours have pressed on my temples, drain from the air, and an inky tropical blackness floods the sky. Daytime sounds fade: the diesel snarl of the big yellow Caterpillars that gnaw on rock down on the river, the twang of carpenters' hammers driving nails into hardwood studs of the new camp buildings. The twilight forest shrieks with insects, and the clatter of suppertime dishes and subdued servants' conversations comes from the kitchens. Then as the last light seeps from the sky, the insect chorus suddenly grows silent. The power-mower sound of our camp generator floats irregularly from a distant hill, and it is night. Closer by, the gecko lizards, whose insatiable appetite for mosquitoes makes them welcome guests on the walls and ceilings of my bungalow, cry *dook-ay* in Thai. I like the lizards, but there is too much biology between us for empathy; I wonder but cannot understand the function of that absurd cry, or if it has no function, whether it corresponds to some need or satisfaction.

A few hundred yards from my quarters is the camp club,

evening hangout for the Thai engineers and foremen, and from it the sounds of occasional shouts or jeers celebrating or protesting some triumph at the billiards table reach my porch. Sometimes after dinner, I walk down to the club to watch the game and talk with the Thai engineers, although gradually I have come to feel that my presence brings an uncomfortable sense of formality to the otherwise relaxed atmosphere. On my arrival tonight, Mr. Suwak, our chief mechanical engineer, leaps from his table, where he can usually be found haranguing some unfortunate subordinate. "Meester Meehan," he shouts in his wild, cracked voice. He is a small man, an inflamed spirit occupying, rather temporarily I think, a sallow and almost fleshless body ravaged by alcohol. His eyes are like the micaceous portholes through which one observes the inner conflagration of a furnace, but his manner is exaggerated and clownish, and he is given to paroxysms of soundless laughter. His general procedure is to stumble about, ordering whiskey for me, falling over chairs in his efforts to align one, all the time delivering speeches of elegant welcome of the type usually made for visiting dignitaries at airports. "I must tell you, Meester Meehan, that you come from a wonderful country, the United States of America, one which we in Thailand admire very, very much, but—and you will have to forgive my English and my manners—during the time that I visited there, in Denver, Colorado, I was unaccustomed to eating eggs for breakfast [this is not a Thai custom] and when I would try, I would throw up." This last he would shout, then throw himself into a pantomime of mirth. Mr. Suwak, much to the detriment of public safety at the camp, was a fancier of guns, and occasionally late at night, I shuddered to hear the boom of a heavy-caliber rifle from the direction of his quarters, across the road from mine. Luay told me, perhaps as a joke, that he was shooting mice, but once, after being away from the project for a few days, I returned to find a bullet hole in the wall of my bedroom. I think him a dangerous man.

But perhaps Mr. Suwak is less mad tonight than usual, I think, seeking company and accepting the invitation to sit at his table. He reminds me, after ordering both of us double whiskeys, that we do have some very admirable and wonderful things in my country.

"Like what?" I ask.

"Like very cheap cars, and Disneyland," he shouts, banging his fist on the table with a ferocity that makes me wonder how his birdlike bones will survive the impact.

Sometimes I talk with Mr. Rangsee, the assistant project engineer. He is a different type, gentle and quiet, polite. He seems afraid of me. Rangsee spent a year and a half in Vicksburg, Mississippi, training in a government exchange program.

"I felt very bad," he tells me, his lips trembling. "They thought that I was a Negro. No one spoke to me all the time I was there. I stayed in a room at a lady's house, and when she saw me, she was angry that the government had not told her I was a Negro, and she would not speak to me."

Or Nicom, who has replaced Vicha as my counterpart engineer. Nicom is drinking rice whiskey again, I notice. Last week, he began to vomit blood one night. "Perhaps you are drinking too much whiskey," I suggest to him. He sits at the metal table, fear in his eyes, a loaded nickel-plated .38 revolver stuck heavily into the waistband of his trousers.

"Do you know about delay poison?" he asks me, in a tight voice. "Delay poison" is probably a direct translation of some Thai idiom, I think, riffling through my conversations with Nirand. I tell him I have heard of such a thing— a poison that some enemy puts in your food. It has no effect for several days, then strikes suddenly. Nicom has fixed on this fairytale as an explanation of his illness.

"I must be careful at all times," he says grimly.

But most nights when I am at the project, I do not go to the club but sit alone in my camp reading *Time* magazine. I smoke Chinese cigars and sip beer. The cold bottle beads with moisture, and as the label grows soft and wet, I peel it with my thumbnail. I have a short-wave radio, a Grundig, infallible and ugly in its brown plastic case, on which I hear the Voice of America, with its thoughtful ten o'clock presentations of news and presentations of U.S. culture. The White House says that the war in Vietnam may end in 1965. Diem is overthrown by military coup. President Sukarno appoints himself premier. Prince Sihanouk severs ties with the United States, despite his fond memories of playing volleyball with U.S. military advisers. Louis Armstrong talks about how he learned to play the trumpet in New Orleans. U.S. chemists devise a rat poison that is harmless to children. The program drones on, like a drunk in a bar. I suppose that I am the only person on the Korat plateau listening.

Sometimes I switch to Radio Peking, which provides tiresome diatribe on American imperialism delivered by a Chinese with an English accent, who always sounds indignant and hurt, on the verge of tears. Recently he has started to complain about the Russians too. Chinese students returning from Moscow, where they have been seriously abused by anti-Leninist elements, are given a tumultuous welcome by the Chinese people.

I write letters home on crinkly blue aerogram paper, which, when dropped into the slot at the post office in Pak Tong Chai, will miraculously fall to Massachusetts, seven thousand miles beneath my feet, in less than a week.

A coil of incense, which is supposed to keep the mosquitoes away, is burning on my table, but the mosquitoes are getting bolder, or hungrier, and I can feel the gentle corona of whirling air that each carries with it as it reconnoiters the inviting terrain of my bare legs.

Outside in the blackness, the innocent and securely repetitive sounds of daytime have gone, given way to silence, broken only by occasional accidental rustlings or forgetful chirps of small creatures who have momentarily forgotten that night has arrived.

Down by the river, the big trees have been cleared and the houses moved from Bu Hua Chan. I have a wire from Bangkok telling me that our new fleet of yellow dozers and scrapers has arrived, so that within the month we will begin to excavate in the borrow pits and start on the dam. In the field behind my bungalow the elephants sway silently as they sleep. They are no longer needed now, and one morning soon I will awake to find them gone.

The Pangal River flows in one of two dozen grand, glacially carved valleys that drain the fluted rampart of the Cordillera de Los Andes in central Chile. The upper reaches of the Pangal, where the milky glacial meltwaters of the Rio Blanco join it, support no permanent habitation, but during the Chilean summer, shepherds with their flocks of gray sheep and mongrel collies range about the patches of meadow that occupy parts of the valley floor and the scattered benches high on the canyon walls. From time to time an explorer traveling on the jeep trail that wanders up the valley will come upon a pair of Chilean cowboys, somber, middle-aged men with ponchos and lariats and broad-rimmed hats, the same *huasos* described by Darwin when he came here a century and a half ago on the *Beagle*. One misty morning on this same road I met a band of outlandish gypsies, leering seductively, as if from a dream, across half a dozen centuries.

The gravelly, boulder-strewn floor of the Pangal Valley is mostly flat and varies in width from a hundred yards to as much as a mile. In its wider stretches the icy waters of the river split into several braided channels that sweep restlessly across the valley floor. From place to place topographic narrows force the Pangal into a single channel that, during the crisp cool autumn days of June, diminishes to the width of a city street and is, in the early part of the day, fordable in a jeep. The flow of the river picks up in the afternoon and early evening when the hot sun accelerates the melting of glaciers ten miles upstream. It is said that the Braden Copper Company, in an ingenious but reportedly futile attempt to stimulate the Pangal's autumn flow, once persuaded the Chilean Air Force to bomb its sparkling white glaciers with bags of black soot.

Here, where the Blanco joins the Pangal and where I lived in a cabin for three months, the valley is a mile above sea level. From the front door of Casa Kaiser, as my little cabin was called, I looked out over an apple orchard, flaming red and yellow in the afternoon autumn sun, across the valley floor at steep rock walls rising abruptly in blue shadow two thousand feet above me; beyond that, another mile up, the peaks of various minor mountains spouted mares' tails of blown snow.

The cordillera is a restless land, said by geologists to be rising into the thin air at a rate of an inch a year. The ice-shattered and wind-blasted volcanic rocks of the mountain peaks lie uncomfortably at these altitudes of two or three miles, and anyone who has spent time in these parts knows well the late afternoon rattle and smack of shards of andesite rock falling from high on the sun-thawed canyon walls. Every few mornings on returning to some part of the valley where I had worked the previous day, I would come upon a sight that would prickle the back of my neck, a slab of rock the size of a Toyota freshly cratered five feet into the river gravels. I was pleased never to have witnessed the more spectacular display that occurs in this canyon every decade or two, when the mountains leap during a great earthquake. It is then, during the fearsome three minutes that these temblors are said to last (time enough for three *credos* according to one terrified fifteenth-century traveler), that entire mountain tops, football stadiums full of rock and ice, tear loose from the heights and fall a mile through clear air to crash onto the valley floor below.

Just such an avalanche had occurred sometime in the past few thousand years about a mile upstream from my cabin, choking the valley floor with a pile of shattered rock several hundred feet deep. At the time of its occurrence this would have presumably interfered with the regular flow of the river, perhaps acting as a natural dam, backing up a lake behind it. Indeed the presence of some old buried mudflats upstream suggested that there must have been a large lake above this avalanche at one time. Whenever or however such a lake may have existed, it had disappeared in prehistoric times when the river had since cut a narrow ravine through the rockslide.

Now this particular physiographic fluke, known locally as the Escofina, the rasp, on account of the ragged profile that the rock pile presented to the sheepherders who climbed over it on their seasonal migrations up the valley, had been of great interest to engineers of the Braden Copper Company for nearly half a century, ever since the days when William Braden began to develop the world's largest underground copper mine, the El Teniente mine, over in the next valley. Transformation of the greenish rock ore into tawny ingots of pure copper requires a steady supply of electric power, of the kind that can be generated most cheaply by diverting nearby alpine rivers into wooden pipes and flumes and then dropping the water at some convenient place through steel penstocks into a hydroelectric generating plant. Much of the power for the mine was generated from water diverted from the Pangal a mile below the Escofina, at the small settlement and pipeline intake known as Bocatoma Pangal. But from the engineers' point of view, nature's cooperation in this enterprise was disappointingly halfhearted. What they wanted was a steady supply of power, which the Pangal, with its spring floods and late summer and fall droughts, perversely failed to provide. If only the spring floodwaters could be stored in a big lake, with a spigot on the bottom, then a uniformly abundant, year-round supply of water and power could be assured. And here at the Escofina, there had once been such a lake, provided by nature free of charge; all that was needed now was a little repair job, easily accomplished by diking the breach in the natural dam with watertight earth and rock embankment.

And so it came about over the years that in an oak file cabinet at Braden's somber offices on Calle Bandera in Santiago a manila file grew fat with accretions of memoranda and reports, calculations of costs and benefits, and fold-softened blueprints (made from drawings elegantly inked on tracing linen and stored in map cases on a different floor of the building) showing the proposed Rio Pangal dam, the earth embankment filling the gap in the Escofina, the spillway and the concrete pipe and valve—the spigot—that would be built beneath it. To "firm up," as we say, the power supply from the Rio Pangal.

Enthusiasm for the project arrived with each new general manager and swelled with the price of copper, but no dam had ever been built at this seemingly ideal site on the Pangal. One can imagine, each decade or so, Braden's chief engineer, gathered with his assistants and eminent consultants on that sterile forlorn hill of loose pinkish rock shards, his blue eyes peering through steel-rimmed glasses at the foaming waters of the Pangal rushing through the ravine. He looks upstream, imagining the huge lake. But what he sees is that the valley is filled with porous gravel and boulders, perhaps a thousand feet deep. And the Escofina, the natural dam that needs only to be plugged, is but a pile of loose rock shards. The chief engineer turns to his staff and consultants. "Can any of you assure me that this reservoir will hold water, that it will even fill in the first place?" Silence. The men are silhouetted against the overcast sky; the pinstriped legs of their dark suits flap in the wind. The river makes a monotonous hollow roar, indifferently dissipating its energy to the rock and the air.

The last such evaluation had been fifteen years before when Braden had hired Kaiser Engineers, one of the big international dam-building firms. Kaiser's project manager, a retired army colonel, had attacked the leakage problem by digging a deep shaft through the upstream side of the Escofina in an attempt to find something more watertight than the loose rock fragments that were scattered over the ground surface. He had not; the shaft filled with water as fast as they could pump it out, which said a lot about the watertightness of the rockslide. Kaiser's report was vague, uncertain. The price of copper was down. The project was scrubbed.

But now again in the mid-1960s Chile's political schizophrenia was flaring up, and there were pressures from the Left for expropriation of the American-owned mines. A compromise plan was developed: Chile would buy a majority of ownership from the Americans, but the Americans would reinvest their income from the sale into expansion of the mine's production facilities, including developing more electrical power. Once again the file containing a half-century of dreams and schemes for the Rio Pangal was pulled from its dusty oak cabinet and a U.S.-based consulting firm specializing in dam design hired to reconsider the feasibility of the project.

I arrived for the first time in Santiago on a fine autumn day in June 1967. The air was soft and smoky, and sunlight filtered through groves of gum trees in dusty slanted columns. Schoolchildren standing by the side of the road watched our jeep pass.

During the past year I had lived a gypsy life on three continents. In the last few days I had driven from Boston to San Francisco in a Volkswagen and in the last twelve hours survived night passage from Los Angeles, latitude thirty-four north, to Santiago, latitude thirty-four south, in an aluminum tube with one hundred fifty other Braniff passengers. The Peruvian stewardesses, upper-class girls, sulked and smoked cigarettes, peevishly filled our orifices with food and earphones. A baby cried through the night.

Off our left wing, behind the black ragged wall of the Andes, the early morning sky filled with frosty pink.

Profound fatigue, geographical dislocation, and fevers bring on an exquisite impressionability in me. Paradigms and points of view fall away with a clatter, significance is rinsed from the landscape, the mind grows soft and waxy. So it was when I arrived in Santiago, I entered an enchanted city. It was as if I had returned to some half-forgotten town in south central Europe and that it was not 1967 but 1937.

My recollections of that day are stored on brittle celluloid reels that project flickering sepia scenes. The damp streets are crowded with walking men. Some are pale, short, and pudgy, with timid eyes and black straight hair slicked down. Others are lean and stringy, with white curly hair and gold smiles. They wear dark wool suits, brown cardigans, worn black shoes with dusty toe caps. From time to time I see insectile encounters between pairs of them, in front of the Café Haiti; they bow slightly, shake hands lightly, touch each others arms and shoulders, talking politics.

"Ah, todos bien, a Dios gracias, Don Ricardo. Supo lo que paso en el Senado?"

"No. Se agarraron a punetazos los Senadores?"

I am in the lobby of the Carrera Hotel. Across the street, in the Plaza des Armas, there is a change of guard. The soldiers look like uncles; they wear heavy wool coats. I have a sore throat. Chilean aspirin are stacked discs in a tube. The hotel is musty and faded, and the elevator disgorges a frail and elderly man in a dark suit, escorting an enormous red-haired prostitute. They both look very happy.

I eat lunch with Federico Schmidt, my contact at Braden and companion-to-be for several weeks on the Rio Pangal. We eat white fish in the hotel dining room, its ambience fit for a fund-raising banquet for Mussolini. The waiters wear greasy tuxedos. The older waiters are obsequious in their dealings with the clientele but humiliate the younger waiters.

Federico is a middle-aged German-Chilean, an engineer and Braden career man. He is precise and formal in his speech. He touches a grayish napkin to his lips after sipping white wine, then refolds the napkin carefully.

After lunch we walk down Calle Bandera. The tawny afternoon light filters through a thick haze. Federico changes ten dollars for seventy *escudos* in a tobacco shop. "Con el dolar por las nubes la situacion hay que cambiarle como se pueda," he says to the storekeeper. Federico is getting seven escudos for the dollar, which is "in the clouds." Six years later, when I return to Chile, it will be not seven but seven hundred.

We walk to Braden's offices. They are done in off-white linoleum, with steel desks. I meet the assistant general manager, a Swiss-Chilean. Señor Mache smokes a cigar and has *Playboy* sent by airmail from the States. I pretend I think that Miami is a wonderful place and give him a fifth of Chivas Regal. Afterward Federico explains the problem of the Rio Pangal dam to me in his cold office. The main concern seems to be the geology of the dam foundation; whether the reservoir will actually hold water is the key question. My assignment will be to supervise certain exploratory works that are due to begin the following week: the drilling of some well holes and excavation of trenches with a bulldozer. I will prepare suitable field records showing progress and results of this and other exploratory work that I consider necessary for design of a dam. Federico will assist me by providing liaison with the contractors who have been retained to perform the exploration. Our field headquarters will be in a small cabin sited in the Pangal valley. Federico's office looks out into an alley. At 4:30 it is getting dark, and rain streaks the dirty windows.

The next day Federico and I drove in a gray Braden jeep from Santiago, sixty miles south to the city of Rancagua, where we stocked up on provisions: bacon and *chuletas* from the *carniceria* (open only two days a week), freshly baked rolls from the *panaderia*, canned beans and a wheel of white cheese from the little *almacen*. I thought at first that the proprietors of these various establishments were relatives or old friends of Federico, for with each of them he exchanged lengthy and elegantly phrased well wishes. But I began to see that that was just Federico's manner, even with people he did not care for; a proper greeting for him was a minor work of art.

Federico at that time was perhaps forty-eight, a trim, neatly if somewhat shabbily attired middle-level civil engineer with fifteen years of service in the Braden Copper Company. His slightly stiff and deliberate manner was offset by blue eyes that were open and friendly, and he had a quick wit and a humor uncharacteristic of his culture. He had an exceptionally good technical mind, and he loved engineering as a subject of philosophical conversation, although never once in the weeks we spent together did I see him apply this talent for the benefit of his employer or any other practical end. "We say of Mister Schmidt," one of the Chilean engineers once told me, flashing a broad gold smile to emphasize the convivial side of what was partially a hostile judgment, "he is a very fine professor."

Federico made elaborate apologies for the inadequacy of his English, pointing out by way of explanation that he had never been in a country where English was the native tongue, or in any other country besides Chile, for that matter. In fact he spoke English with only an exotic hint of middle European accent and, drawing from a marvelously rich English vocabulary, crafted phrases and sentences with the kind of original skill and attention to detail that seemed to belong to some lost age. The slight imperfections of nuance and anachronistic sentence structure brought a special bouquet to Federico's conversation, as certain impurities do to wine, and made me smile, or want to smile, at everything he said.

After we stocked up on provisions at Rancagua, our next stop, an hour's drive up into the Andean foothills, halfway between Rancagua and Braden's El Teniente Mine, was the little town of Coya, where the company's main executive offices were situated. Here was the elaborately landscaped home of the much-talked-about but rarely seen general manager of the mine; the company golf course; the trim complex of offices, built on the banks of a river around a little quadrangle, like a pre–World War II army headquarters. Coya, with its convivial dining hall, its corps of pretty secretaries, and pleasant apartments, was a last outpost of sophisticated social life. From there, twenty miles up one valley, perched on the steep side of a barren rock slope, belching clouds of sulphurous smoke and discharging a torrent of leaden sludge, was the El Teniente mine. Twenty miles up the adjacent canyon was the small settlement of Bocatoma Pangal, and above that the mile-high Pangal valley where Federico and I were to live.

We stopped at Coya for several hours to discuss this and that matter with various officials and subofficials. Federico loved making elaborate arrangements, stopping in to chat with various secretaries, allotting at least ten minutes to each conversation. In those days I considered this on-the-job socializing a great waste of time; I was very impatient to "get things done" in the pre-1960s American way, and this impatience of mine combined with Federico's inefficiency and procrastination eventually created strain between us. But it is doubtless true that I would not have made as many friends at Coya without Federico. Many times during the weeks that we spent on the Pangal, when the work did not go well and we became discouraged, he would persuade us that the lack of some minor provision— some Band-Aids, for example—was sufficient reason for us to drive that afternoon an hour and a half down the rocky road to Coya, then having procured or not procured the Band-Aids, to stay on for tea, and perhaps a party afterward. Then midnight would find us, mellow with Chilean cabernet and the smiles of young women, zig-zagging in our jeep up the switchbacks to our cabin on the river.

The Braden office in Santiago had made arrangements for a Chilean contractor to provide a drilling rig and support crew and a bulldozer and operator to carry out our exploration program at the dam site. Federico and I spent the first week of our stay on the Pangal selecting sites for drill holes and test pits and supervising the emplacement of our equipment. Our drill rig was an ancient well-drilling machine of the percussion type, which advances a hole in hard, bouldery ground by the simple method of repeatedly dropping a half-ton pointed steel rod and then flushing out the broken rock fragments with pumped water. The machine could not obtain diagnostic cores of the rock, so it was necessary for me to evaluate the subsurface conditions by observing the behavior and sound of the dropped steel bit (whether it clanged or thudded, for example) and by examining the dirty water flushed from the hole.

Each morning and afternoon Federico and I would visit the drilling site. Federico would chat with the four men who stacked casing and pipe, welded and maintained the sputtering gasoline engine that powered the rig. They were an odd gentlemanly group, unlike the usual crew of peasants and roughnecks, successors to nineteenth-century sailors, who usually work the drilling business. One of them, whom Federico addressed as Señor Villalobos, had fine hands and a noble head. He did not seem eager to talk, but when pressed by Federico he would comment laconically in such a way that suggested he was a well-educated man, which led Federico and me to speculate over coffee that evening that he was the disgraced son of some distinguished Spanish family. While Federico chatted with the drillers, I would examine the material flushed from the drill hole and make notes on the progress of the drilling. My daily reports, recorded in a surveyor's notebook, began to look the same; a disappointing meter or less of progress each day, penetrating the same shattered andesite rock that I could see all around me on the ground surface.

I had sited this hole on the upstream side of the Escofina, where the rock slide blended into a large alluvial fan that debouched from a side canyon. If the Escofina had once dammed the valley, I reasoned that I should find evidence of something watertight within it. But each day, each new meter, reduced my enthusiasm. Directly on the other side of the river was the abandoned shaft excavated by the Kaiser engineers years before. All they had found was loose, shattered rock, and it looked as if my work was going to yield the same result.

Besides the drill rig, our other piece of exploratory equipment was a battered yellow Caterpillar D-8 bulldozer, which was meant to cut crude roads for moving our drill rig and also to excavate trenches in the ground for my inspection and analysis. This old machine suffered from a kind of chronic rheumatism, and for two-thirds of the time during the next three months it was down for maintenance of one kind or another. I used to marvel at the patience and cleverness of its operator, a bearish Chilean who singlehandedly would disassemble major pieces of this huge machine while it sulked immobile on some windy rock knoll or half-submerged in an icy pool of water. Then he would rebuild the worn or broken parts in a small welding shop at Bocatoma Pangal and put the whole thing back together, only to have some other major component go a few hours after the miraculous restoration. But the machine did remain operating long enough for me to accomplish one important objective, which was to excavate some deep trenches in the meadow upstream of the Esco-fina. It was there, after digging through ten feet or so of boulders, that I encountered a peculiar thing—a layer of fine brown silt, of the type that one finds deposited only in a large lake, a lake like the one we proposed to create by damming the Escofina.

Progress was slow in these enterprises, but after a while my frustration at the inability of our equipment to probe the immensities of that barren landscape subsided some-what, and Federico and I settled into a comfortable daily routine. Federico in his German way enjoyed scheduled activities, and he took satisfaction in the smooth operation of our household, the preparation of elaborate meals, the setting of table with the eclectic collection of china and silverware, the methodical cleaning and polishing of cookware and dishes, the final rinse of all eating utensils with boiling water.

I am inclined to slip into carelessness in my living hab-its and, what is worse, to subject my daily routine to engi-neering analysis, in that MIT way. (Should I just butter my bread with my fork, thereby saving the effort of washing the knife later on?) But during the first few weeks at Casa Kaiser, before Federico returned to Santiago and I came to live alone there, the cabin became a glowing outpost of civilization in that bleached and stony country. As the au-tumn days shrank and damp mists came down from the glaciers above, warmth and monastical order prevailed within Casa Kaiser, sit-ups were done just so before retir-ing, soiled laundry folded neatly, the pleasures of food and conversation enhanced by ritualization. "Now Don Ri-cardo," Federico would say as we took our after-dinner coffee, cooked in a battered saucepan with exactly half an eggshell to settle the grounds, "you were saying in regard to these 'hippies,' this morning on the Escofina, before cir-cumstances interrupted us . . . "

After we had been drilling and trenching the dam site for a month, there was no more arranging for Federico to do, so he returned to Santiago to attend to other matters, leaving me alone at Casa Kaiser. By this time, progress on the drilling was decelerating as the hole got deeper; I fell into a lethargic routine, and the days began to seem very much the same. Every morning at nine and then again in the afternoon, I would drive my jeep down from the little hill on which Casa Kaiser was sited, through the apple orchard, now turning rust and brown and frosty in the early winter mornings, then across the bouldery floor of the valley, over the pink shattered rock of the Escofina, as sterile and timeless as the surface of the moon, then back onto the big alluvial fan that debouched into the valley upstream of the rock slide. Now in late July the sun sank lower in the sky each day, and the canyon was mostly in shade. In the afternoon leaden clouds began to crowd against the sharp bare ridges above, and a harsh wind rippled the black icy pools of water on the rocky flats. Out on the fan, our drill crew gathered piles of bleached dead brush, which they ignited with gasoline and which the wind-whipped flames reduced to cold white ash in minutes. Señor Villalobos, looking even more the part of an exiled nobleman, wrapped himself in a ragged blanket of coarse llama wool, squatted next to the dying ashes, and squinted up at the darkening sky to the west. "Nieve," he said to me late one afternoon, as the air suddenly whirled with fine dry snowflakes. Within a few minutes the light faded, the air became still, and I could not see more than twenty yards for the snow, falling heavily and silently just as it does in one of those round glass balls containing a miniature alpine village, which when inverted, fill with slowly falling white flakes. That afternoon I made the last entry under CD-1, churn drill hole number one, recording progress for the six hours of drilling that day. It was a brief entry: "Fifteen centimeters today. More boulders."

It snowed hard and the cabin shook in the wind like a man with fever all night, and when I got up early the next morning the storm seemed to have settled into the valley for the day, and I knew that I would not be going anywhere for a while. I boiled coffee in a saucepan, camp-style with an eggshell, sliced a piece of bacon and fried it with an egg, and cleaned my plate with half of a hardened Chilean roll. Then I methodically washed and dried the dishes, after the way of Federico, and studied Spanish, the preterit and imperfect of *hacer*, until the heat of the cabin and the food made me sleepy and I went back to bed, sleeping heavily and dreamlessly for two hours. Then I got up again and sat at the oilcloth-covered kitchen table, reading through the thick Braden file on the Pangal dam project. Two years before, a famous engineer had come to visit the site, and he wrote a letter report advising the company on various technical points related to construction of the dam. In one part of the letter he described how the boulders could be raked from the alluvial fan upstream from the dam, so that the fan could be covered with a blanket of compacted silt, to stop water from leaking through the dam foundation. I had never seen such a thing actually done, and I tried to imagine it, carefully modeling the process in my mind, a Caterpillar D-9 pushing a rake, tines as thick as my thigh, steel on rock squealing and sparking, the dusty ozone smell of shattered andesite, the boulders torn from nests they had occupied for years, perhaps ten or ten thousand, no one knew. Some of the boulders were as big as Plymouth Rock. Lacking experience, I tried to imagine how they would be moved, whether the blade of the D-8 would do it, or whether they'd "mudcap" them, the way they had taught us in the army, where you put a couple of sticks of dynamite on the boulder and then cover the dynamite with mud, so that the blast is somehow directed back into the boulder and not into the air in a way that I did not understand. At this point I paused to try to work out the Newtonian mechanics of the process, the role of the mudcap's mass in the delivery of energy to the boulder, filling a page with little diagrams and equations from freshman physics at MIT, then later trying to remember whether $F = ma$ was merely a definition of F or whether it actually told something useful.

From time to time during the day, between these labors and amusements, I would get up and look out of the window at the snow. Sometimes it fell thickly and quietly. At other times, when the gusts gathered to blow hard until Casa Kaiser creaked and shuddered, it hissed and scratched icily on the window glass and whirled horizontally across the white plain, clouds of it lifting from the drifts until there was no longer any clear physical distinction between the snow that was falling and that which had already fallen.

Standing there alone, at once liberated and overwhelmed by forces as beatific and indifferent as the sea, I felt carefree and happy. It seemed to me that the day was a realization of a Waldenesque fantasy I had built twenty years before, during a prepubescent hunting and fishing phase, the year that I had subscribed to both *Outdoor Life* and *Field and Stream*. I remembered how I used to imagine myself a lone trapper, without a dog to keep me company, tending my line of traps in the winter. I even harassed my mother into a solemn agreement that she would permit me to buy a lever action .22 and take a year off from school when I was fifteen so that I could test my ability to survive, in Wisconsin I believe it was, with my rifle, one box of rimfire cartridges, a fur-lined parka, and a tobacco tin containing wooden matches (tips waterproofed by dipping them in candlewax), a needle and thread, and some chicken bouillon cubes. Somehow this venture had been forgotten when I was fifteen; but when I arose the next morning, thirteen years behind schedule, I was there. Looking out from the window of my cabin I saw a virginal blue sky, the pale pink and cream and slate-rocks of the peaks two thousand feet above me, and in the valley, an icy blue sea of snow. Snowbound, I thought; and then to savor the pleasure that words carried, I said it aloud.

Later that morning Señor Vargas, chief foreman at the nearby headworks, beat a path through the drifts between his camp at the Bocatoma and Casa Kaiser. He pounded on my door, which I could open only a few inches because of the snow piled against it. Señor Vargas and I always had a few friendly words for each other. "Nieve mucha," he shouted clearly, waving his arms and looking all around at the white substance that had so pervasively invaded our lives, displaying his brown teeth in a mighty smile. "Si, blanco," I said enthusiastically. It amused me to think of Señor Vargas telling his wife about our conversations.

We got down to practical matters. Does one open the road to Coya, I asked. No, not open, said Vargas. Much snow. Cerillos, can't be passed, much snow. The Cerillos were the steep switchbacks that lay midway between Bocatoma Pangal and Coya. Of course I knew the road was impassable—the average snowfall had been about four feet—but with Vargas's arrival the spell had been broken and I was already looking ahead, impatient to escape this confinement. I tried unsuccessfully to extract a prediction from Vargas. I guess he just did not know, and it did not matter to him anyway; he wasn't going anywhere. Somehow I always expected the Chilean locals were omniscient, that they knew the population of towns, the distances between any two points, the Spanish names of rare birds and trees, and the weather for the next thirty days. Vargas left me with a cheerful smile, saying he would have his men shovel the snow away from the door of the cabin.

It was a long day. That afternoon the phone rang, long-short-long, long-short-long. It was Paula, the chief engineer's secretary, calling from Coya. "How are you," she asked. "Do you have enough to eat?" She did not know what the schedule was on clearing roads but would find out and call back tomorrow. I asked her to tell Driver, the chief engineer, that we were shut down and couldn't even get to the drill rig. I would come down as soon as the road was cleared.

That evening I wrote my weekly report. Citing the lack of progress under the best of conditions, the worsening weather that had now shut us down entirely, and the inadequacy of our equipment, I concluded that it was useless to go on with the current exploratory program. Our three months of effort had yielded little or nothing in the way of useful information. We still had no idea whether a reservoir at the site would ever fill with water. If anything, the results of all of our work indicated that it would not.

That night I skimmed along the deep pool of sleep. Minutes expanded grotesquely into quarter-hours. Next morning I went out to walk, to clear my head, but a biting wind drove me back into the cabin. I drank too much coffee and let the breakfast dishes sit dirty in the sink all day. I found myself believing that washing socks was a social custom that lost any meaning to a hermit. So it went for the next two days. I began to sleep a lot—twelve, fourteen hours a day. I became dyspeptic and began to hate Chile.

I had lived overseas for more than three years now, and the romance of it was fast dissipating. I wanted to go home, to read *The New York Times* and see Perry Mason on television, eat pizza and listen to "Blowing in the Wind," perhaps dare to try a hula hoop. And yet I knew that I could never go back to all that; it was 1967 and home was not the same anymore. The month before I had come to Chile, tens of thousands had gathered in Sheep Meadow and Golden Gate Park to celebrate love. Timothy Leary had spoken to the multitudes: "Turn on to the scene, tune in to what's happening, and drop out—of high school, college, grade school . . . follow me, the hard way."

What would I do back home? Perhaps I would go back to school, study for a doctorate. That would keep me occupied for a couple of years. Then I would be thirty. Surely then I could find some purpose, something I really enjoyed.

The three days since the storm had been unseasonably warm; the valley dazzled and gurgled in the hot sun and great icicles grew on the eaves of Casa Kaiser. By the fourth day the snow had mostly melted on the muddy jeep trail that went from my cabin to the Escofina, and that afternoon I drove out to the rock slide and climbed up from the road to its highest part. There I sat on a sun-warmed rock and looked down on the dam site spread out below me, the jointed rock on the canyon walls, the bouldery fans scarred by debris flow tracks and partially covered by the pink shattered rock of the Escofina. Upstream the braided alluvial gravels were pocked from place to place by bulldozer trenches, exposing delicately layered, brown lake silts that lay like chocolate cake beneath the river gravels. Sitting on that rock I tried to reconstruct the series of geologic events, catastrophic or imperceptibly gradual, that had built this landscape. But there were essential exhibits and explanations missing from the museum diorama that I built and manipulated in my mind. I began with the observation that the layers, or varves, of brown silt exposed in my trenches could have been deposited only in the still water of a lake, one that froze every year. There were hundreds of these layers; therefore the lake must have existed for hundreds of years. If nature had already built a reservoir here, we should only have to repair the breach in the rock slide dam to restore nature's lake for our own purposes. And yet in all of the site explorations on the Escofina, Kaiser's deep shaft and our trenches and the churn drill hole, we found only loose broken rock. When I tried to pump water out of Kaiser's shaft, it rushed back in as fast as the fifty gallons a minute we could pump. That was at least a number, and when I took that number and translated it mathematically into a leakage rate for a reservoir, the rate was such that the reservoir would never fill. But that's not what happened. It had filled before. Something prevented the water from leaking out. But what? Then I would look over that forlorn landscape again, at the porous gravel and the loose rocky debris of the fans and the rock slide. Whatever it was, it must be hidden from view.

There was a secret waiting to be uncovered; this stony valley was teasing me with clues, and I could feel a new and subtle kind of tension capture me as I worked to find the answer. Across the river I could see the churn drill, the steel shaft with its pointed bit hanging impotent above the dark hole in the earth. So far, it was twelve meters, and nothing but the same sterile rock that I could see all around me.

Back at the cabin that evening, I imagined that I had found a watertight foundation and drew sketches of various kinds of dams. A concrete arch or buttress dam probably would not work. I would build it of earth and rock. It would be a two hundred-foot high dike, the central part of which, the core, would be made from leak-proof silt or clay. Where would I get that material? It would be hard to find. Perhaps the meadow upstream from the Escofina. There was a thin surficial layer of silt there, deposited by little man-made irrigation canals maintained for decades by the shepherds. But it was a thin layer, and if we dug too deep with heavy construction equipment, we would come up with a lot of gravel mixed in with it. How about the varved lake silts? Possible, but they might be too wet, on the soupy side, hard to compact properly and impossible to dry up here. Anyway I had no idea how widespread these deposits were; I'd hit them in only two of my trenches. How about a plain rockfill embankment, then build a concrete slab on the upstream side to seal it from water? Let's say a slab eighteen inches thick, a seven-sack concrete mix. Later I would ask Federico about the cost of a sack of concrete in Chile, and then I would calculate some rough quantities and costs. But how would a structure like that stand up in an earthquake? The concrete would be prone to crack. Not so good. Expensive too. Start again. How about a hydraulic fill dam? Build it by hosing dirt and rock off the side of the mountain above the site, diking it into pools so the muddy fraction settles out in the center of the dam, slowly adding layer on layer, the way they built them in the California Sierra during the gold mining days? Now that's an exciting concept. Simple. Really simple. Pumps and hoses, wooden flumes, no imported machinery with the big Chilean import duty. But there are earthquake problems again; hydraulic fills are notoriously unstable in a big shake.

So I went on, making little sketches and cost estimates, building concepts and then knocking them down.

A friend of mine who is not an engineer once asked me a question that was so simple that I was unable to answer it. "What is it," he inquired one day after I had been talking about my work for half an hour or so, "that you engineers actually do?" Well I guess what we do is to fiddle with problems the way I did during my snowbound days on the Pangal. I had a client, and the client had a need—in this case more water at Bocatoma Pangal in the late summer and fall, when the natural river flow was low. So I thought about that need a lot, making sure that I stated it in the simplest way possible, without any built-in, assumed solutions (he needs water in the penstock, more water going into the pipe at Bocatoma Pangal, but not necessarily more water in the river upstream of the intake; maybe we could pump extra water out of the ground rather than try to get more river flow). Then when I got the need boiled down to the simplest possible terms I started playing with what I had to work with—men and machines and materials and time—translating these resources into 1965 dollars for purposes of comparing solutions. For instance, I might have a permanent staff of ten thousand men standing next to the river year round, and in the flood season they would scoop five-gallon buckets of water from the river and carry them to high ground, five minutes each trip, so I would have two thousand buckets full each minute, that's ten thousand gallons a minute, more than twenty cubic feet per second, just about what we need in the dry season, when they dump it back in. So we have to build a town of sixty thousand people (three shifts, wives and children) and arrange to haul in bread and brilliantine and brown cardigans. Then I would translate the cost of all that back into 1965 *escudos*, figuring how much of an outlay I would need to build and pay all the expenses of that operation for the next, say, fifty years. Then if the economics looked at all attractive on that first crude pass at the problem, I would go back and begin to worry about some of the details: would the buckets leak and how often would they have to be replaced and how much water would evaporate while the buckets were sitting out there all summer? Then there would be questions of reliability. Suppose all these people went on strike or caught the flu? And safety; do you want to build a town in this valley, with rocks the size of railroad locomotives sailing down out of the sky every few months? That's just how they wiped out a school full of children in Wales a few years back. And what about the sheepherders and the rainbow trout that thrive in the water and the fact that this is a fine-looking place now; does anyone—the Chilean government, for example— have anything to say about that? So there are certain rules to the game, besides the basic rule of meeting the need with the least costly solution, and these rules become the design criteria.

In this way I would go over the problem again and again,

then sleep on it, and think about it while I was lying in bed at dawn. The aesthetic is simplicity, and I would imagine fantastic solutions: I would study the problem for a year, then issue a one-page report: "At precisely 2460.5 meters S 80°05′16.2″ E of Benchmark R45, high on the canyon wall above the Escofina, there is a large crevasse in the rock. Two sticks of Dupont 60% dynamite or equivalent should be placed at this location precisely as shown on my sketch, primed, and the valley below cleared. This explosive charge will detach an already unstable block of rock approximately 4 million cubic meters in volume. According to my calculations, the resulting rock slide will plug the gap in the Escofina, creating a reservoir between 10 and 15 million cubic meters in volume."

Engineers are thought to be practical people, and yet the core of engineering contains elements of idealism that are absent in other professions, medicine and law. I mean idealism in the original platonic sense, the belief that somewhere is the perfect design, the perfect dam. We'll never see it, and we'll certainly not come close to building it, but we know enough about it to be forever seeking it.

For example, we know that its construction and subsequent function represent a domestication rather than a perversion of nature. This implies collaboration; we seek out and seize any opportunity provided by advantageous topographic and geologic conditions in the valley. There is the narrow place in the canyon caused in this case by the fortuitous rock slide, the expanse of flat valley upstream, with its deposits of silt available for our use. Like rich mineral deposits, these features and resources are accidents of landscape, anomalies, places where nature's entropic decline toward dull uniformity has been locally perturbed by some external force. The strategy of site selection is a matter of recognizing and exploiting flaws in the natural fabric. We are like the tree that seeks out the soil-filled crack on the rock wall of the canyon, but we perform our search with deliberate intelligence rather than by scattering a million seeds over a square mile of landscape.

This notion of collaboration applies similarly to the engineering use of materials in the fabricated elements of our design. For example, concrete, like eggshell, is watertight, strong in compression, weak in tension, plastic at the time of fabrication, brittle later on. We create from it elements of our design that our intuition, confirmed by mathematical idealization, shows us are in pure compression— hence the double-arch concrete dam. But hereon the Pangal, the dam will rest not on the hard rock but more elastic alluvial soils, which when loaded, respond in such a way that accidental tensions will creep into our idealized force field. And concrete will not take tension; it lacks flexibility. Whatever material is used to build the dam, it must be cheap, watertight, strong, and flexible, and we quickly find that there is no material in our catalog that will meet these four criteria. Lacking a material with these virtues, we split them and consider a composite design. Wet clay is cleap, watertight, and flexible, but its strength is low. Rock fill is cheap and strong but not watertight. Our solution then is similar to nature's ingenious but trial-and-error solution to the problem of the human spine, where a plastic vertebral nucleus is enclosed with a case of bone; we contain a core of clay within shells of rock.

Having developed a crude design concept, we move on to encounter the host of subsidiary problems that must be solved to polish the rough product. Among these is the question of how much is enough. For example, design traditions dictate that an earth dam must not be allowed to be overtopped by floodwaters, because it may wash away. Hence we need a special spillway to carry floodwaters around the dam when the reservoir is full. How big does the spillway have to be? Is it enough to design it to carry the flood that comes along once a century? Once a millennium? Should we also armor the top of the earth dam with rock to prevent it from eroding away just in case the spillway is not big enough (say, if an unpredictable avalanche falls into the reservoir and causes a big wave)? Questions of component adequacy and redundancy. Can we squeeze a bonus out of redundancy, as nature gives us stereoscopic vision with our spare eye?

It was there on the Pangal, as the valley thawed and I waited for the road down to Coya to be cleared, that I learned the pleasure in it, in this design. For in the end it does not differ from any other art, the satisfaction in making a clay bowl or a painting or writing a sentence or a symphony. First the concept, the trial efforts, the crude shape of a good solution, then refinements, balance, and polish until the final arrangement sings with deceiving simplicity and stuns with accuracy of effect. For the first time I was able to experience technology not as the stepchild of science (which is, after all, impotent) but as an art. A few years before at MIT, I learned to be an engineer. Now, here in Chile, in this odd backwater of civilization, I was learning to be an *ingeniero*.

One afternoon a week after the storm, I heard the diesel snort of a machine, and when I walked down to the little settlement of Bocatoma Pangal, there was a strange Caterpillar D-6 on the road, up from Coya, sitting there smugly, the little engine that could. That meant the road was open, and that afternoon I drove my jeep down to Coya and went to see chief engineer Driver, explaining to him my views on the futility of continuing the exploration program. Driver seemed indecisive on the matter. He was an American expatriate, one who had settled comfortably into a life of relaxed prosperity. Now in the last year or so, chaos was beating at the door. The Chileans had taken over 51 percent of Braden's ownership, and there was talk of a large expansion program engineered by an outside U.S. design-construction firm. A substantial step forward in the Rio

Blanco investigation, which was his responsibility, would improve the security of his position. But when I described our drilling program, he had no choice but to agree with me and asked that I go back to the site for a few more days to supervise removal of the drill rig and bulldozer from the valley.

The next day found me again on the Pangal, directing my drill crew to build a wooden cover to seal the now-to-be-abandoned CD-1 hole. By afternoon the drill rig had been dismantled and hauled down to Bocatoma Pangal ready for pickup by a flatbed truck.

The last piece of equipment to leave the upper valley was our ancient bulldozer. After we backfilled the trenches in the meadow, there was no further work for this machine, and I dismissed the operator. It was a warm, bright afternoon in the valley and I stood on the Escofina, stripped down to a T-shirt watching the snow-fed Pangal running strong below in the ravine and the bulldozer crawling like a yellow beetle on the bank of the river. Part of the road was submerged, and to pass that part the tractor edged up the side of the steep, rocky slope, gouging a narrow trail for itself with its blade, sending the pink rock cascading down the steep slope into the churning water below. When I was standing there, perhaps one hundred yards away, I noticed something peculiar about the ragged path that the machine tracked behind it; where the blade had cut into the slope the deepest, the earth was an odd, dark gray, slatey color. As the tracks of the bulldozer and the edge of its blade cut into this slope, the harsh squeal of steel on rock stopped, and all I could hear was the rush of water and the tired sound of the diesel engine. I clambered down the broken rock scree to the dozer, waving at the operator to stop. There, beneath a two-foot layer of loose rock that covered the slope, was exposed a compact mixture of gravel and sand with abundant portions of gray clay. The material was damp but not wet, except where water from melting snow was running down from the freshly cut slope into puddles on the flat surface of the pad cut by the dozer blade. There the water stood, without percolating into the ground, and there it remained all that day and the next day when I came back to look again. Suddenly everything was clear. All of our exploration, all of the past exploration too, had concentrated on finding a buried impervious barrier within the rock slide, but on its upstream side. And all anyone had found, to a depth of a hundred feet, was loose, porous rock. But here, at the downstream side of the slide, where the dozer blade had cut into it, the pink loose rock of the Escofina was only a deceptive veneer, covering a compact, clayey, watertight deposit of glacial till. It was a terminal moraine that had once dammed the Pangal valley, backing a large lake upstream, no different from the countless morainal lakes found in glaciated valleys throughout the world. The rock slide had not dammed the valley after all; perhaps it had even occurred long after the moraine dam had breached and the lake drained away. But it had thoroughly covered the natural moraine dam with loose rock, effectively disguising its shape and character and even its very existence. Now for the first time I had more than a fantasy foundation to work with.

The next day I went back down to Coya and explained the discovery to chief engineer Driver. Personal matters required that I return to California. But with his enthusiastic endorsement, I sent a cable to California requesting that someone be sent down at once to replace me and continue the exploration program as I directed. What I had at this point was still a hypothesis, not a proof, and the next step in the scientific method was to test that hypothesis by excavating a pit or shaft in some other part of the Escofina, which I reckoned should also have a core of watertight moraine.

Two days later I met my replacement, a construction

man named Larry Hatch, at Pudahuel Airport and brought him to the site. I had arranged for an experienced well-digging crew to arrive at the Pangal the next day, having decided that the previous drilling operation was too slow and its results too difficult to interpret, and I gave Hatch and his crew instructions to begin excavating by hand a vertical shaft from a high spot on the Escofina. "You're going to go through a lot of loose rock like you see on the ground here," I told them, "but sooner or later I think you are going to hit something different, like this." I showed them the distinctively gray, watertight clay exposed in the bulldozer cut. The shaft would advance at about a meter a day, and the walls would have to be cemented as they went, to prevent caving. But each day Hatch was to record what he saw and take a Polaroid picture of the freshly exposed shaft walls, forwarding this information to me every week. The next day I said good-bye to Señor Vargas and Driver and the girls at Coya, and to Federico, and caught the evening flight out of Santiago to Los Angeles. As we climbed into the blue-black clear sky I looked down at the valleys, now shaded and filling with mist, and the frosted rampart of the Andes, incised every few miles by rivers the size of the Pangal. I was glad to be leaving but excited by what I was beginning to feel was my project.

Back in Calfironia I went out on a new assignment, the site investigation and design of a dam to be located in a dry, bony landscape, on the edge of the southern California desert. Here I found an entirely different display of geologic oddities, and I was soon absorbed, with my geologist partner, in a collaborative effort to untangle a history of moving earthquake faults and rivers that were no more. During the evenings, I sat on a plastic turquoise chair at a wobbly table in a motel next to a retirement mobile home park, devising schemes to recreate, for the benefit of the Metropolitan Water District of Southern California, a lake that had dried up seven thousand years before. One day a pack of ten coyotes ran through the site, and two Mexicans walked over a hill and asked me the direction to Los Angeles.

But each week, with the arrival of a battered package containing a handwritten report and a collection of Polaroid photographs, I was reminded again of the Pangal.

For a while, as I had suspected, our shaft penetrated nothing but loose, broken rock. But then one Friday afternoon about two months later, Hatch's report contained some important news. They had been excavating at a depth of about seventy feet when he suddenly noticed one afternoon that the buckets of earth that the crew pulled from the hole looked different. Climbing down the shaft on a rope ladder, he took a Polaroid picture, enclosed. I looked at the picture. The top half showed jagged shards of rock identical to the previous two dozen photos, obviously rockslide debris. But the lower half of the picture showed a gray, homogeneously clayey soil. "Water is running into the hole from the top of this layer," said Hatch's report. "This is a big problem because the water sits in the hole and we have to pump it out twice a day." For the first time I was confident that we had a dam site on the Rio Pangal.

During the next few days I planned a program of continuing field investigations: I would probe the Escofina and the river channel in a few dozen more places to trace the top of the watertight moraine, test the silts from the meadow upstream, where I planned to get the impervious core material for my dam. And I began to work up some more detailed plans for my dam embankment and to take on some special safety problems—earthquakes and the dreaded possibility of a major avalanche falling into the full reservoir. Then one day there was a new dramatic report from Chile; Braden's engineering department had been reorganized. Driver and all of the other foreigners were leaving the now-Chilean-controlled company; the American mining interests were pulling out or being pushed out. There would no longer be any possibility of a major investment in a new dam. This I learned in a cordial personal letter written in a fine hand with an old-fashioned fountain pen. The same writer also sent a tersely worded official letter typed on stationery of Sociedad Minera El Teniente, S.A., the new name of the former Braden mining enterprise. It gave me formal notice to stop all further work on the Pangal dam project and instructed me to submit a final billing for my services. Further work would not be authorized in the foreseeable future, it went on to say. The letter was posted at Coya and signed by the new chief engineer, Federico B. Schmidt.

The Pangal still runs free above the Bocatoma, and I suppose that with the soft international copper market and the inefficiencies and labor unrest that have plagued Chile's now fully nationalized mining industry for the past decade, there is little need for more power at the El Teniente mine. But irrigation water is in short supply in Chile and in 1974 when I was again in the Chilean Andes looking at a potential dam project on the Rio Pardones, the next canyon south of the Pangal, I was shown a report by some Chilean engineers suggesting a new irrigation project that would require construction of a dam at my site on the Pangal. It was a proven site, according to the optimistic report. I wouldn't go so far as to say that, but it won't surprise me either if one of these years a dam is finally built on the Escofina.

For my part, fulfillment of the project would have provided the satisfaction that comes when ideas are materialized. But I am not greatly disappointed that the dam was never built. It was enough for me to have advanced the concept from a vague desire to a design, albeit one needing further confirmation by field exploration. What I learned there on the Pangal is the spirit and potency of design, and that is something that has stuck with me since, even through the dark romantic age of environmentalism when most of us were driven from the lush jungle of engineering creativity to the arid highlands of engineering analysis.

In one sense, a piece of the Pangal concept has become realized and lives today transplanted in the upper part of a dam recently completed in southern California. During the course of my work on the Pangal, the possibility of another earthquake-triggered, Escofina-like avalanche had worried me, and as an extra precaution against overtopping and erosive failure of the dam due to the mammoth splash it would create falling into the reservoir, I had planned to build a special erosion-resistant, armored zone in the upper part of the Pangal dam. This zone was to consist of large boulders raked from the Escofina. The measure is a redundancy and therefore a departure from design orthodoxy. But I transplanted the idea to my southern California dam anyway as an extra precaution against its overtopping during an earthquake. I was pleased, not to say surprised, when it was later given some minor recognition, within professional circles, as an original design innovation. It mattered little to me that another engineer, whose need for attention exceeded my own, claimed credit for it; I believe he was punished for that arrogance by having to attend a chicken and creamed pea award dinner at a seedy banquet room in Houston, Texas. For my part, I was glad to have missed the affair, for I was occupied in some detective work at a new dam site.

The Pacific Northwest city that I shall here call Micaville is situated on the fringe of a douglas fir forest at the foot of a range of granite mountains. The houses on the steep streets above Main Street are of turn-of-the-century, West Coast, Victorian style and overlook in the middle distance an agricultural valley that on a spring day, is checkered in greens and umbers and mottled with the blown shadows of puffy cumulus clouds. It is a bucolic landscape, romantic in the English sense; it struck me on first sight as a fit setting for the stories of Peter Rabbit. Micaville itself is a local cultural center, site of a small college and a summer music festival. It is also a timber town, supporting enterprises that produce broom handles, as well as all of the wooden bases for the most popular national brand of mouse trap. (This last item of information I learned from Daniel Dempsey, Micaville's waterworks superintendent.)

The city derives its name from the muscovite mica that sparkles in the sugary granite exposed in road cuts in the forest above town. Shiny flakes of the same mineral streak the sand in the bed of Mica Creek, a freshet that emerges from a fern-filled grotto above town, flowing through the city in a spacious landscaped park. The park is the former retirement estate of a prominent midwestern industrialist, designed after the natural style of nineteenth-century English gardens. During July afternoons, the park provides a setting in which visiting distinguished musicians, local teenagers, itinerant motorcyclists, and conservatively dressed tourists sit on rustic benches by the swan pond or stroll peaceably on the shady, hand-raked sand paths along Mica Creek, like beasts in a coniferous Eden.

The granitic sand of Mica Creek, with its bright micaceous lustre, creamy white grains of feldspar, and carbon black streaks of organic litter, is of professional interest to me; in fact, a jar of it sits before me on my desk. I have been spending some time this evening studying a series of colored aerial photographs that show the origins of this sand, which I maintain are the bright yellowish-white cuts and fills of the Forest Service road that winds prominently through the green hills above town. Also shown on the photographs midway between the road-scarred forest and the town is the blue-black sapphire of Mullens Reservoir, which captures and stores Mica Creek's snowmelt waters for summer use by the city. Bob Curtis, who is Micaville's director of public works, and I took these photographs one gusty day this spring while flying a shuddering Cessna 172 five hundred feet above the treetops.

Micaville's city management currently is engaged in a bureaucratic dispute with the U.S. Forest Service, the agency that owns the forest above town. A few years ago, the Forest Service opened these lands by building some roads in the Mica Creek watershed and permitting some patches of clear-cut logging. The next winter's rains washed one hundred thousand tons of sand off the disturbed, friable hillsides in the clear-cut areas and down into the creek bed. Then the following winter, a flash flood caused the accumulated sand to ooze down the clogged creek like wet concrete, ending up in and partially filling Mullens Reservoir. The city maintains that this kind of disturbance of the ground or vegetation in the watershed eventually will fill the reservoir, clog its water mains, destroying Micaville's only source of water. The Forest Service, which is accountable to the timber industry as well as to the city, is attempting to compromise in the usual bureaucratic manner. I have been hired as a consultant to the city to document on their behalf the causes of the sediment problem. Bob Curtis, city public works director and my client, would like to see the watershed locked up as a preserve as the best means of protecting the water supply.

But now in the late hours of this California summer evening I have pushed aside the photographs and my tentative report outline, written on white bond with a fine fountain pen in jet black permanent ink. These and other research materials—Kent's *The Effect of Wildfire on Sediment Yield in the Idaho Batholith* (1976) and Mullens's *A Proposal to Enhance the Water Supply of Micaville* (1910)—being the most recent and most ancient—I gather into a tidy pile.

I slide open the glass door that leads to the central atrium of the apartment building in which I live. The sight and aroma of the courtyard, the fluorescent eye of its swimming pool, the stage-lit orange and lemon trees, hibiscus, and monstrous Australian tree ferns, resonate across a thirty-year octave of memory with boyhood fantasies of warmer places, conceived amid the dry brush grays and browns and frosty yellows of a New England boyhood. It is one o'clock in the morning.

I am not yet ready to write my report on this matter. There is a part of my mind, the logical, analytical part, that I imagine to be the mental analogue to my fountain pen (a heavy black Mont Blanc, its nib now gleaming gold and clean in the bluish light of my desk lamp). But there is another part of the mind that is the paper, sometimes smooth and white, but other times like tonight, a rough and fibrous stuff on which the fine line of writing blurs to unwanted shapes. That is the way it is tonight, after several false starts. And so I think it is not yet time to write a report but to reflect some upon this assignment to exhume from memory certain events and people and, having subjected them to examination, propitiate their ghosts. Only then can I return to this rational inquiry into the matter of sediment in Mica Creek.

Sometimes the students at the university where I teach

or the recent graduates who work for me in my practice express frustration at their inability to write technical reports with what they incorrectly imagine is the effortlessness of those of us who have been at it for years. I try to explain to them that it cannot be done until certain shadowy processes have run their course, that I too spend many hours sitting before blank pieces of paper, staring at the patch of steel blue dawn that appears in my window each morning, waiting for congealed mental juices to begin to flow. I explain my idea that a report is not so much a document meant to be read as it is a sort of trial to which one subjects oneself, to determine whether a well-formed professional opinion has run its full course of gestation. I do not attempt to explain to them that I believe there are certain queer, unspeakable processes that occur in the interim. That is a matter they will have to discover for themselves.

Bob Curtis's responsibilities include supervision of the City Water Department. Bob is a civil engineer, in his early forties, an easy-going man who smiles at visitors when they walk into his small office. I couldn't ask for a better client than Bob. Everyone in town seems to like him. Bob is so agreeable that he even read *Small Is Beautiful* at the suggestion of Brian Moore, Micaville's young city manager. So I was surprised one day when Bob showed a side of himself that seemed a little vindictive, even though I suppose that it is what some people would expect from a director of public works.

I had arrived at Micaville's little airport on the ten o'clock United flight, rented a car, and drove downtown. It was a cloudy morning, threatening more snow; it had snowed the night before, and the fir forest above town was powdered white. The snow had mostly melted downtown, although some of the cars still wore chains. I parked next to Mica Creek, in the park, and met Bob at his office on the second floor of the city administrative building. Micaville does not have much of a city hall, and all its services are jammed into an old commercial building that faces onto the town square, across from the park and the fountain from which you can get a drink of natural mineral water piped from springs up in the hills. Mica Creek flows through the center of town, in the park; this morning its waters were running silvery black with the micaceous mud washed down from the reservoir. The purpose of my trip here was to see the now almost empty reservoir, which was being cleaned of its annual accumulation of sediment.

Bob and I had lunch at a small restaurant with a Franklin stove, run by a beatific group of young people, the sort of place that seems to be springing up everywhere—nuts and alfalfa sprouts, apple juice, and yogurt—not the sort of restaurant most directors of public works would go to. But Bob seemed to like it, and he even smiled when he read his horoscope on the menu. It was a tasty lunch, and we talked about flying. Both Bob and I are private pilots. I have a lot of clients in positions like Bob's, and often I have to fake my conversation when I'm with them. This is part of the consulting business, and it's hard work for me. But I felt easy with Bob. We thought alike. Or so I thought.

After lunch, we drove in Bob's city pickup, a four-wheel drive, on the slushy, unpaved road that climbs up the canyon above the city to Mullens Reservoir. The trees and ferns and rocks on the canyon walls carried translucent slices and towers of snow. The overcast was breaking up, and the canyon began to sparkle in the slanting rays of the early afternoon sun. When we stopped the truck to look at the creek halfway up to the water treatment plant, it was beautiful and very quiet.

A few minutes later, we drove the truck on to the treatment plant, it was there that Bob told me the story about the girl and the dog that surprised me and made me a little wary of him for the rest of my assignment.

At the treatment plant, the road narrows and squeezes between a concrete wall and the natural granite canyon wall; anyone walking up the road to Mullens Reservoir and the unposted public watershed lands above must walk through this narrow pass. The treatment plant operator owns a dog, a big german shepherd, with a fierce hatred of all strangers. As we approached the plant, the dog attacked the wheels of our truck, snarling and drooling. The dog was a monster, with glazed eyes, a creature from a Greek myth.

"That dog sure helps to keep the hippies out of the watershed," Bob said in a pleasant voice, smiling. "Last year the operator told me about one hippie couple who were trying to walk up the road to the watershed."

Bob used the word hippie loosely. I knew he was talking about someone like my daughter, with Elton John records and patched jeans, a boy friend with a beard. Or perhaps the couple who ran the restaurant we had left a few minutes ago.

"The dog waited until they got about here," he went on. "Then he went after them, and the girl started to run. That's the wrong thing to do with that dog; the guy stood still, and the dog went right by him, after the girl. He kept nipping at her as she ran. She was wearing one of those homemade halter tops that you just tie on, and as she ran, it loosened up. She tried to grab at it, but the dog was right on her, and it fell off, and there she was, running along, the dog snapping at her. The operator said it was a funny sight to see." He laughed and shook his head. I stared at him, astonished, unable to reconcile what I had just heard with the pleasantly philosophical man with whom I had spent the last hour. Did I fail to see the humor in the story? Would the men at Bu Hua Chan, the village in Thailand where I had once lived, get the joke?

Bob is a professional man, in some sense a gentleman and

an officer. Dan Dempsey is his first sergeant. Dan has worked for the city for twenty-five years, starting as a laborer, and in a few years he'll retire from his present position as water superintendent.

One day Dan and I were sitting next to a creek in a conifer forest in the hills two thousand feet above Micaville in the Forest Service lands. Spring had come, and the sky was laundered fresh and pale, like old denim. We were eating ham sandwiches that Dan had brought from town in paper sacks.

"I once bought a 1932 Ford," Dan was saying to me, beginning to tell his tenth or fifteenth anecdote of the day. Not so much *tell* it; Dan was *bawling* it out, in his first sergeant way.

"One Sunday morning," he went on, "the car was sitting in the driveway, when I look out and there's a man standing next to it. So I put on some clothes and go out.

" 'What can I do for you?' I say to him.

" 'How much you want for that car?' he says.

" 'Well,' I tell him, 'that car ain't for sale.'

" 'Oh yes it is,' he tells me.

"Well it turns out that he's some rich fella, buys old cars and fixes 'em up, and he's willing to pay a pretty good price for that old car.

" 'Mister,' I say, 'If I didn't have a dime in the world, and you offered me five hundred dollars for that car, and it wasn't worth twenty-five dollars, I wouldn't take it.'

" 'Well,' he says to me, 'You don't have to be so indignant about it.' "

Dan reproduced the dialogue with mirth, in the Irish way. He nodded his head, pleased with the recollection of this adamant refusal. No rich son of a bitch tells him what to do.

Dan and I get along pretty well. Like Bob, he doesn't care for hippies, but I can understand that more in Dan's case, as a matter of class. Dan does not like people who are

impressed with their own importance either. He spends a
lot of energy talking about his past wars, all of which he
claims to have won. He wears a yellow cap with the name
and emblem of a seed company on it. He has a manner of
speaking that is nominally belligerent, although I really
don't pay much attention to that side of him but somehow
just read between the lines. What is it that makes me know
that Dan is, for all his talk, a kindly man?

The suburban city where I live is urbane and wealthy; its
concerns are of the kind one reads about in the *New
Yorker*'s "Talk of the Town." Days or weeks sometimes pass
before I see a pregnant woman, and there are strict laws
about parking recreational vehicles and boats in drive-
ways. Living here, I forget about what I think of as the
enlisted men and their wives, the young couples I once
knew. Until I stay in a place like Micaville for a few days,
then I begin to notice them on the streets—the young men
with T-shirts and jeans, tentatively experimental beards, a
furtive rural look; the girls, wispy or overweight, pale with
caved-in tired eyes, smoking cigarettes.

Take Dwayne Hayes, for example, of 323½ River Street,
Micaville, recently laid off as a yardman at the mouse trap
factory. He is waiting until the city and Forest Service resolve the matter of timber harvesting above town. He
might get a job again if they begin cutting. Dwayne appears
to me as a rather unhealthy model for a 1944 Norman
Rockwell painting of freshly recruited draftees. It is amaz-
ing how different I feel about one of those *Saturday Evening Post* covers when its characters begin to move, and
those endearing characteristics—rabbity teeth, hairy moles,
clogged pores, bad posture—spring into discomforting reality, and I feel the corona of animal heat that each body
carries, and I hold my breath.

Dwayne was like that—pale, fuzzy patches of blond
beard, smelling sourly of clothes unwashed for many days,

standing, confused and uncertain, at the intersection of Lincoln and Seventh streets.

Pamela Mathews, also of 323½ River Street, I took to be his common-law wife. She was short and fat, with a kindly, maternal face and cheap plastic-framed glasses. She was frightened and needed someone to hold her; lacking that, she was rubbing her elbow absently and holding a healthy pink infant, who appeared to be asleep.

We were just becoming acquainted with Dwayne and Pamela and their child, having driven our Hertz Monte Carlo, cadmium red with white interior, 300 horsepower, smartly into the side of their aged Falcon, muddy white, hubcaps missing, blanket spread over the tattered remnants of upholstery. My working associate, Dale Swanson, architect of this messy accident, was explaining the whole thing to Officer Eddie Fagan, badge 12, of the Micaville Police Department. Dale was assuaging his guilt by wrecking his legal position. "It was my fault," he was saying. "I never even saw the red light until Dick here"—he nodded at me—"said 'you're going through a red light.'" That was a good thing for me to have said at that time, I reflected. It caused Dale to brake an instant after I spoke; as a Mormon nondrinker and nonsmoker, Dale has superb reaction time. As it was, we hit the rear door of the Falcon as it swerved desperately across our path. Had I not spoken, we would have continued through the intersection at thirty-five miles per hour, and the Falcon would have smashed broadside, at fifty miles per hour, into my side of the car. So I was feeling very lucky, but at the same time wincing at Dale's ingenuous cooperation with Eddie Fagan.

But as we attempted to behave as gentlemen and good citizens, feeling relieved and thankful that the infant was uninjured, we blundered into foolish positions, displaying the weakness of the ingenuously thankful. Dale was our spokesman, and Officer Fagan who, although on the verge, I felt, of advising us otherwise, had settled into his official investigative role. "Do you really want me to investigate this accident?" he had inquired of Dale. The proper answer to this question, I see now, and Dwayne Hayes saw at the time, was, "No, officer, there are only a couple of minor scratches on the cars, no one hurt, and we can handle the whole thing through our insurance agents."

But in the end, Dale, who wanted to see everything done right, talked too much, and all Eddie Fagan contributed to the situation (and what else could he have contributed, when I think about it) was a citation to Dale for running the red light, along with a quasi-judicial finding that Dale caused the accident. To Dwayne, the victim of it all, Fagan gave advice that his driver's licence would be suspended because his vehicle—actually registered in Pamela's name— was uninsured. Perhaps, I reflected, it takes a poor man to understand when to keep the authorities out of things.

There is a point to this; each professional assignment is a replaying of some old drama, each new character a phantom from the past.

Only now in the dawn outside my window as the sky fills with a liquid gray light, I see more clearly as if on a stage backlit, the silhouettes of my characters, Curtis, Dempsey, and Hayes, dressed in the green fatigues of the U.S. Army, and I remember now how unprepared I felt, fifteen years ago, to join them. I too wore the uniform, but I never could escape the feeling that it was as an imposter, an outsider. There was among these men a certain complicity; it seemed to me they each had found their pattern, and this discovery, more than the symbols of rank on their uniforms, created a bond, a masculine pact from which I felt excluded.

It had turned out just that way one summer night in 1962, at three o'clock in the morning, when O'Brien, my sergeant of the guard, told me he had been raised in Micaville. I had heard of the city but had never been to or known anyone from that part of the country.

I was a second lieutenant then, costumed in forest green with this curiously real .45 strapped to my waist by a canvas belt. I had just returned in a jeep from an inspection tour of the North Post of Fort Belvoir, Viginia. My guards were scattered over ten square miles of motor pools, rail depots, and construction equipment demonstration areas. The inspection had gone well; none of the guards had been asleep, and each had satisfactorily answered my questions about general orders and how to report fires. It was a fine night, with a bright moon that gave the parade ground a frosty glow and made strong black shadows beneath the oak trees and between the barracks. Having been out in the fresh air for an hour, I was wide awake. Sergeant O'Brien was sitting in the guard room reading a coverless copy of True magazine when I got back. When he flipped off the light, lit a cigarette, and asked me how the inspection had gone, I took that as reason enough to sit down and talk for a while.

O'Brien was a draftee, twenty-five years old or so, a couple of years older than I. He carried a compact 180 pounds erectly and with an easy confidence. O'Brien had a way with the men that really impressed me.

"You, Wilson, I want you to take the lieutenant's jeep over to the motor pool, check the air pressure in the left rear tire; that's the left rear tire, got it? And get your ass back here in ten minutes."

And Wilson, a pale southern boy, a sullen troublemaker in my experience, far from resenting such arrogance, was even openly pleased with this attention, would have wagged his tail if he had one, and presented some little joke to O'Brien like an offering of thanks. Later in the evening, when Wilson was giving instructions to his squad, I could hear a lot of O'Brien in his style of speaking.

Anyway O'Brien and I were the only ones awake at three

in the morning, and we sat talking and smoking Lucky Strikes. I asked him whether he was planning to stay in the army.

"No sir, I am getting short, two months to go, then I'm out of this army," he said, a little mockery in the "sir." I had a month to go myself.

"Where're you going when you get out?" I asked. I had in mind that O'Brien knew exactly where he was going, that he was going to do something clever, to run something. As for myself, I wasn't sure; I had been writing letters, well composed and finicky, to engineering firms, hoping to get a job offer by return mail. So far nothing very promising had come back.

"As soon as I get out of here, lieutenant, I'm going to Oregon."

We were sitting on bunk beds, and our cigarettes glowed in the half-darkness. It was a hot night; the air smelled of men's sleep, and the doors and windows of the orderly room were open. In the distance, I could hear the first kitchen sounds from the 588th, the slam of car doors, clatter of pots, men whistling and coughing, half a mile down the road. A slanted shaft of street light cut through our smoke and reflected dully on the smooth walnut stocks and black oily steel of the rack of M1s in the middle of the room. It was a time to be sort of philosophical, even sentimental, for Dana Andrews to push his helmet back on his head and talk about this big farm in Montana he's going to buy when the war is over.

"I've never been to Oregon." I said. "I'd like to go there, though; from the pictures I've seen, it must be a nice place. Big forests of fir trees with clear streams full of salmon and cliffs overlooking the Pacific." I was from the East Coast, and that was exactly what came to mind when I thought the word *Oregon*: an illuminated beer advertisement.

"Well, lieutenant, that may be what comes to your mind about Oregon, but it's not what comes to my mind." I had that sick feeling of the hopelessly trapped; O'Brien had set this up, and now he was moving in for the kill. "What comes to my mind about Oregon is junk. That's where the action is in Oregon. When I go there, I'm going to start a junkyard."

O'Brien stubbed out his cigarette, stood up, and yawned.

"Well, lieutenant, we've got forty-five minutes before the next guard, and I think I'll get a little shuteye."

We technological experts pride ourselves in arriving at findings on such matters as the logging of Mica Creek through the use of pure reason; we think that our training immunizes us to feeling. And yet "the head is always the dupe of the heart," said the Duc de la Rochefoucauld.

I have observed here that one can quickly find, even in a small town visited for the first time, a great deal that is very familiar, and with that familiarity natural objectivity fades and is overwhelmed by attitude. In this way, the preparation of a successful technical report is not a recording of the workings of a purely reasonable intelligence but rather a part of the process by which the mind laboriously frees itself from the constraints of feeling. It is this that has been the most difficult part of the task, and only now, that work done, with the world awakening outside my door, can I return to my outline and that first blank page.

When I was nineteen, they put me in charge of a major hydro-electric project, an attempt to recreate the Tennessee Valley Authority in Haiti's Artibonite valley. My first task was to check out Haiti's sociopolitical background. I soon found that matters had hardly improved there since the chief of state, Christophe, shot himself with a silver bullet and the queen disposed of the royal corpse in a bucket of lye. The place was crawling with President Duvalier's secret police, the dreaded Ton-ton Macoute. Superstition was flourishing. Soil erosion was ruining the coffee bushes. Privately I had my doubts whether the Artibonite project would really bring Haiti into the space age. But my job was to get the project designed, not ask questions. So I rounded up a team of three other engineers to work out the details.

Tim McCreary was responsible for the big dam. He decided, on what basis I do not know, to design it as a concrete gravity dam, and he made a study, based on an article he read in *Engineering News-Record*, of the economics of pumpcrete. Rafael Garcia, a sophisticated Cuban who chain smoked filter-tipped cigarettes and privately thought we were crazy to be wasting time on such a forlorn place as Haiti when we could be drinking rum and Coca-Cola at Club Latino, was supposed to design the irrigation system. He didn't show up at our second or any subsequent coordination meetings, so in the end there wasn't any irrigation system, and the dam just spewed water into the jungle.

Frank Moss handled the power distribution system, with his usual enthusiasm. Frank was from Florida. He had actually been to Haiti once with his older brother. "What's it really like there?" I would ask him, the way my daughter would a few years later ask me to repeat a favorite story. Frank would smile and say that it was really interesting and then go on to tell me some story or other. I made vague plans to go to Haiti myself some day, maybe some Easter vacation.

"Civil Engineering Projects I" was supposed to stimulate our creativity and illustrate practical application of the other subjects we were studying in our sophomore year at MIT, moments of inertia and water-cement ratios and D'Arcy's law. The course was intended to give us the feel of a big civil engineering project. Gordon Williams taught the course. Williams was a traditionalist who believed engineers should be designers and builders, men of action. This World War II view was falling out of fashion in the Civil Engineering Department at MIT in the 1950s. Theory and analysis were back in style. In this Williams sniffed the first whiffs of decay. He soon left MIT, or perhaps was driven into exile. A few years later, in 1963, I ran into him in New York, while walking on Park Avenue during lunch hour. It turned out that we were working for the same consulting firm. MIT was turning out theorists who did not know how to do anything practical, he complained. Perhaps he was right, and that's why in 1979 the Germans provide the design and we provide the cheap labor to build Volkswagen Rabbits.

Our other civil course that fall was "Surveying I," taught by Professor Miller, a rising star of the radical camp. Miller did not much care about traditional surveying matters, the arts of chaining distances and tapping reticle screws. He used "Surveying I" as a front for the doctrines of numerical analysis. Once I asked him why they even still called the course "Surveying," and he said they had to keep the name for the image; otherwise the alumni would complain. Miller taught us a kind of scientific archaeology: how to sift through middens of observational data, separating pottery shards from pebbles, applying such principles as Newton's backward-difference interpolation formula to re-create the ethereal form of the original artifact.

Williams's and Miller's classes were our introduction to civil engineering. Whether by accident or design, they at once made us aware that engineering, like any other art or profession, was a divided house. The split was along traditional lines: Williams, a romantic, promoted design, synthesis, practice; Miller, a classicist, taught us reduction, theory, analysis. At other times, these fundamentally different styles might have coexisted peacefully, perhaps even collaboratively, but by the late 1950s MIT's once-proud Civil Engineering Department was on the skids; only 15 of 650 sophomores in the class of 1961 had elected a civil engineering major, and in accordance with the laws that govern institutions, the threat of extinction had stimulated some revolutionary thinking.

At the time, all of this was much over my head, for that year I was busy discovering myself in the literary sorrows of Goethe and the megalomaniac ravings of Rousseau. It was the Victorian image of civil engineering, the Panama Canal and the Brooklyn Bridge, that attracted me to Building One, with its hall of forgotten engines, designed to perform heroic tests, the bending of iron beams and bursting of concrete cylinders. The progressive side of the business, operations research and systems planning, repelled me, for I saw in it the beginning of a corruption that would lead to a life on commuter trains, working for a big corporation, voting the Eisenhower-Nixon ticket.

That autumn of 1957 was not a happy time for me. When I think back on it now, I remember Wednesday afternoons, sitting in Building One's second floor drafting room. I had two hours to kill before ROTC drill. I wore a brown army officer's uniform, a hand-me-down from the days of Panmunjom. Winter was setting in; dank mists blew down the

Charles and the Cambridge air stank of rubber. At four o'clock I would run across Massachusetts Avenue, against the light, to the sorry armory. Rain was beginning to spatter against the dark windows, and the evening traffic hissed and thickened on Memorial Drive. A map of Haiti was spread out on the drafting table before me. I stared at it, dreaming of voodoo and drums and ripe fruit.

It took Miller's whiz-kid analysts only two or three years to win dominance of MIT's Civil Engineering Department. By the time I graduated in 1961, the traditional design-oriented group had been all but obliterated, and most of us had been converted to the analytical creed. Although I have always admired synthesis over analysis, my talents are inclined more toward the latter than the former, and moreover I am secretly a terrible conformist. So by the time I was a senior, having recovered from my sophomoric eruption of *Sturm und Drang*, I became an enlightened acolyte in the church of ratiocination.

Reason, in its striving to replace religion, borrows many of its forms, and the variety of rational mystique promulgated in MIT's Department of Civil Engineering in the 1950s was no exception. It offered the system general, ruler of heaven and earth. It rediscovered, in the obscure eighteenth-century clergyman Thomas Bayes, a prophet, whose theorems infallibly provided the correct decision on planning picnics, on selecting highway routes, on matters of love and of war. It maintained, in the basement of Building One amid unused, dusty machines, an IBM 650 computer, a faithful and oracular beast that, when fed electricity and punched cardboard, guided us on our way with its million winking eyes, reducing the infuriating complexities of the world to a vast array of zeros and ones.

Was there not in all of this a vague intimation of some great truth, waiting to be discovered? "We're chasing it down," said one of my young professors. He was referring to the analytical solution of some problem in soil mechanics, but the glaze in his eyes and the ripple of excitement that passed through the classroom suggested some more exciting quest than the improvement of retaining walls, a faint promise of ecstasy. Like Pythagoras, we believed we heard in the din of the world around us the siren of the spheres.

Nearly four hundred delegates from a dozen countries attended the Fourth Pan American Conference on Soil Mechanics and Foundation Engineering, held at a large, milky-pink hotel on the isle of Puerto Rico, in June 1971. The three volumes of the conference proceedings contain, among other matters, 138 pages of text and equations dealing with the interpretation of results of the standard penetration test, including one 86-page treatise with a bibliography of no fewer than 353 references, prepared by the prominent and energetic Brazilian engineer, V. F. B. De Mello, during a year he spent as visiting professor at MIT. The standard penetration test (SPT), first invented in the 1930s, is performed to test the firmness of the ground and consists of driving a piece of 2-inch diameter pipe into the soil with a 180-pound hammer, counting the number of blows required to advance the pipe 12 inches. For a while during the 1960s and 1970s, the value of this test was pooh-poohed by many engineers who favored the use of high-technology methods to perform similar evaluations, but now in 1980, perhaps owing to the Marcopolian discovery that the Chinese have been using this untransistorized test all along, the SPT is enjoying the same renaissance of respectability as windmills, herbal teas, and drawstring pants. Only last week I was awarded a contract to conduct several dozen of the tests at the site of a certain dam.

Erudite analysis of pipe hammering is the business of soil mechanics (erdbaumechanic), a new branch of engineering science developed by the Austrian engineer K.

Terzaghi. "The task of soil mechanics," Terzaghi announced, "is to forecast the effect upon soil of a given system of forces and to estimate the pressure of soil against retaining structures." Terzaghi was a product of that oddly fertile mold that formed around the last of the Austrian empire and sprouted such eccentrics as Mahler, Freud, Wittgenstein, and Hitler. He was (to use his own words) one of "those few who were born under a luckier star." Following an appropriately misspent youth of drinking and dueling interspersed with study of philosophy, Terzaghi exiled himself to Turkey in his middle years, and there during the 1920s developed his scientific concepts about soils with the aid of rubber bands and cigar boxes. Upon completion of his work, he compared himself to Darwin, considering both himself and the Englishman midwives of ideas whose time had come.

Terzaghi's principal accomplishment was to extend the nineteenth-century mechanics of steel and concrete to the less cooperative domain of clay and gravel and rock. His method, introduced in engineering circles in the 1930s, was widely acclaimed as a triumph of reason over the caprice of nature, and the image of Terzaghi, if not the man himself, received in engineering circles the honors accorded to a spiritual leader. By the 1970s, soil mechanics had attracted some ten thousand professional specialists, and a complete bibliography of the subject contained some one hundred thousand articles and books. Recently the methods of soil mechanics have been extended further still into such areas as earthquake prediction and estimation of environmental impacts.

Successful ideologies carry the elements of their own destruction. So it was with soil mechanics; the concept became an institution, drawing swarms of government-funded researchers who created baroque but useless variations on the original theme. In practice, the method offered a powerful but dangerous appeal: it seemed to reduce the geological world to a simple and predictable engine. So, inevitably, too bold decisions and imprudently grand designs were founded on this shaky misinterpretation of Terzaghi's purpose. Some catastrophic failures followed. Terzaghi himself saw the danger and issued warnings that went mostly unheeded. "To practice an art successfully," he said, "one must possess the capacity, ascribed to Theodore Roosevelt, for thinking with the hips. In other words, one must be able to arrive at correct conclusions without preceding logical reasoning." Terzaghi's later papers were published in obscure journals. The idea that large dams and nuclear power plants were designed without preceding logical reasoning was not agreeable to their sponsors and owners. To most of us young engineers who burned with the cold fire of rationalism, unlearned in this refined use of the hips, Terzaghi seemed to be backsliding, like Isaac Newton, into cranky mysticism.

For several months I myself had been chasing down the answer to a certain technical problem having to do with the earthquake behavior of a proposed concrete retaining wall, and in late spring of 1971 I prescribed for myself a tax-deductible tropical holiday as a cure for that persistent hum in the wiring that comes when I have been too long at such a task.

My retaining wall, which is plagued even to this day by nonexistence, was a small component of a large construction project visualized by its owner as being a significant part of our way of life in 1990. The project was opposed by a group of wealthy and articulate zealots who believed that their way of life in 1971 was already quite satisfactory and who, moreover, opposed the intrusion of the project into a singularly rustic reach of Pacific coastal landscape. As was customary in the 1970s, the opposition was disguised as a concern for public safety, and in accordance with this method, the protesters had cleverly found in the earthquake safety of the wall a small flaw in the otherwise seamless technological defense of the facility's design. Whether the flaw had any real significance hardly mattered; it offered a focus for lengthy scientific debate, and thereby would serve at least to delay the project and perhaps, in the end, starve it to death.

The owner's chief engineer had retained me as a consultant to assist in resolving the issue. My assignment was to analyze the earthquake behavior of the wall using the most sophisticated analysis I could get my hands on. The idea behind the sophistication was to overwhelm the various regulatory agencies' technical staffs, who would then support the owner's claims against the cynical objections of the "kooks" who opposed the project. The chief engineer, a reluctant Goliath, was under heavy pressure from his company management, to the extent that I felt that somewhere in the basement of his mind his very masculinity was at stake.

I thought I understood the guerrilla tactics of the opposition, recognizing in their preliminary moves what was becoming in the late 1970s a standard strategy of entrapment. I knew that the theoretical basis of the most advanced computer analysis available was out of date and contained some untenable assumptions about the nature of earthquakes. In the real world, as opposed to the idealized analytical world, earthquakes were actually worse, and retaining walls were actually better, than supposed in the analytical model. Hence a simpler, more realistic demonstration of the adequacy of the wall was preferable to the sophisticated (and for me, professionally profitable) house of cards that the chief engineer wanted. I knew that eventually the underlying assumptions of the most sophisticated analytical technique would be subject to scientific, as opposed to engineering, scrutiny and that agreement among scientists of various persuasions would never be attained. Hence the nontechnical commission would be ultimately faced with an unresolved scientific debate and would, accordingly, not approve the project.

I tried to make this point, which was against my own interest, with the chief engineer and, failing that, provided the machiavellian advice that we put on a great but feigned show of defense, then at the last minute simply remove the wall, like a diseased organ, from the project. Or at least rearrange things so that earthquake-induced movement of the wall would be demonstrably innocuous.

The chief engineer was growing impatient. Did I want to do the analysis or not? Perhaps on the basis of a naive faith in the world of codes, or in computers, or perhaps because he had a grasp of the interconnections joining the worlds of words, numbers, and reality that exceeded my own, he had his mind made up in favor of computerized sophistication, and he wanted to get on with it, sparing no expense.

Now months later, my labors were nearly finished, and I had submitted a draft report of the results of my analyses to the chief engineer. I was tired, and the Puerto Rico conference had come along at just the right time. The prospect of a tropical setting, the opportunity to see some old East Coast friends, the extensive buffet of technical papers promised to sooth my nerves.

Alas, Puerto Rico proved a disappointment. I had visited the island a decade before, in the late 1950s, and my recollection of that ripe chunk of limestone, steaming in an azure sea, was accordingly out of date. Now in 1971 the sienna-colored earth, once thick with tangled brush and sugar cane, was checkered with FHA subdivisions, the white tropical mist stained with the exhaust chopped Chevies and channeled Mercs, the rich ecology of ants and squashed toads replaced by damp tourists and irritable hotel employees. I am, moreover, soon enervated by the social strains of conferences; a few days of inept cultivation of potential clients and simultaneous evasion of favor seekers produces in me fungal rot of the spirit.

At seven o'clock on the third evening of the week-long conference, I was standing on the fringe of a cocktail party pretending to inspect a cubic centimeter of cheese that I held mounted on a red plastic toothpick. Everyone else seemed to be engaged in intense conversation. I happened to be reflecting that the various "urgent" telephone messages that had been announced during the conference proceedings were generally for those of my fellow engineers who seemed most desirous and needy of public attention. These thoughts were interrupted by an acquaintance who asked me whether I had seen the note on the conference bulletin board that I was supposed to call my office. This news produced in me two simultaneous neurohormonal squirts—the first, of distress (death in the family? delivery of a summons?). The second was an involuntarily pleasurable little flush, comparable to that thrill that "Miss Joy Golden" doubtless experienced a few minutes ago as her name was called over the public address system here beside the pool of the Beverly Hills Hotel where I am now writing.

Back in my room, I returned the call to one of my partners, a geologist. "I don't understand this stuff," he began, "but I think you should know that one of the client's engineers—a Chandra Bahadur, do you know him? Anyway, he called and said that his contact at the regulatory staff called him and said that they can't consider your analysis because the computer program you used doesn't give the same result as some other program they've used on three other licensing cases. He said it was interesting, but it would be a waste of time for them to even review it. The chief engineer wants to see you, tomorrow if possible, so I said I'd try to get in touch with you."

"Why are you telling me this? I'm supposed to be on vacation."

"Well, I just thought you should know," he said.

Ten minutes later, I placed a call to Air France. The next day at noon, while the rest of the delegates considered the rival taxonomics for materials within pockets of decalcified marl, I was two hours out and four hundred miles west of San Juan, all seats and trays were being returned to their upright position, and the captain was lowering flaps for our approach to Port-au-Prince International Airport. They said that telephone service between Haiti and the United States was "practically impossible."

The Hotel Normandie clings to a steep hill thick with bougainvillea and poinsetta above the town of Petionville, fifteen hundred feet above Port-au-Prince. The evenings are soft and dark and alive, and from the veranda of the hotel the view of the city below and the bay is splendid. Monsieur DeVries, a retired railroad clerk and family friend of the French Canadian proprietor, operates the establishment during the off-season summer, while the owners vacation in Quebec.

My wife and I were the only guests. There was nothing to read and, owing to the steepness of the hill and the heat and darkness, nowhere to walk; we sat on the veranda after dinner drinking coffee. DeVries joined us, offering his bottle of brandy and lighting a Cuban cigar. DeVries was a widower who lived alone during the winter with his routine in a Montreal apartment. He enjoyed a kind of stable, healthy fatness. The humor in his eyes and the black hair sprouting from his ears and nostrils testified to a kind of inner fertility that made me like him at once.

Simone cleared the rosewood table and brought coffee. Simone did everything but cook during the off-season. I guessed she was in her late twenties. She wore a white uniform, which was a pleasing contrast against her soft mocha skin. There was a dark warmth in her glance at DeVries. DeVries considered the Haitians to be uncivilized, but I felt that a part of him was nonetheless at home here.

"They are children," DeVries was saying. "They live for the moment only. I will tell you one example; Simone here, she is perhaps thirty, thirty-five years old? Her husband was a driver at the embassy of Peru. Well, there are certain politicians here that had this problem with the police, the *macoutes*." DeVries lowered his voice and leaned toward us. "You know these fellows, these *macoutes*; perhaps you have seen them at the Cabane Choucoune, the big ones with the sun glasses, sitting in cars. Well, one night two months ago, her husband sent a message to Simone that he would have to leave because the *macoutes* were going to . . . " DeVries ran his finger across his throat. "She has heard nothing of him since, the husband, but last week I asked our other girl, who does the laundry, about the husband of Simone. 'Oh, we do not know where he is, but he is alive,' she told me. 'How do you know that,' I asked her. 'Well, you know Simone's two children; last week, they were lost for two days, and then on Sunday they found them dead, in the sewer.' That was the revenge of the *macoutes* on the husband because he flew away. And Simone, she came to work as usual the next morning; she was quiet for a few days, but now you see, it is only one month since this problem, and she smiles and sings. There will be more children. She goes on living; that is her idea of life, and who can say that is not the better way?"

The last light had seeped from the sky, and the city lights below us twinkled in the smoky heat. Here on the dark hillside the air was thick with odors and voices and the barking of dogs. What was the death of two children? Life went on.

The narrow, concrete-paved road above Petionville in the morning becomes a streambed into which trickle, from red dirt tributaries, women and burros and handpainted camionettes, with names like "Sainte la Lune," laden with bananas and mangoes and burlap sacks of charcoal, streaming down the mountainside into the pool of white haze that fills the bay of Port-au-Prince and the Cul-de-Sac plain below. Jean Paul's sprung Buick Dynaflow, after breakfasting on two dollars worth of Esso, gamely fights its way up the road against the current, its horn bawling. Jean Paul has twelve children and has been a licensed guide for fourteen years. I ask him when his windshield got cracked. He has accentuated the spider-web pattern by painting the cracks vermilion and yellow. He is a cheerful man, but there is something that annoys him about my question. "Very beautiful," he says when I ask him again. The history of the Buick is not part of the regular touristic program.

I speak very distinctly, using only the present tense. We live in San Francisco. En français, La Ville de Saint François, I say. Jean Paul smiles broadly. Yes François, he says. His brother lives in Miami. Very beautiful. We agree to correspond regularly, knowing that nothing will ever come of it. My polyester short-sleeved shirt is becoming translucent with sweat.

The road is notched into the side of the moutain, exposing a core of rotten limestone beneath a rind of dark red soil, that *terra rosa* that gives no rest to the eye in the lower latitudes. Unlike the roads of Puerto Rico—those California freeways set in the unlikely tropical landscape, designed to federal standards on the basis of preconstruction laboratory tests and computerized stability analyses, and constructed with Caterpillar and International Harvester scrapers—this Haitian road is built by hand, its design pragmatic, not rational. How big should this culvert be, where the road crosses this streambed, now dry in the winter? Well, the stream looks smaller than the one that washed out the road in the big rain two years ago, that we fixed with the big pipe. But now we have no big pipe, only much smaller pipe. Perhaps two small pipes; we shall try

that and see. And here, at this corner, where the boys dug
into the hill last week, the soil is deep and very wet from
a spring; surely it will slide down into the road in the first
rain. Any fool can see that. A man takes a few days' vacation from the job, a few days only, and they do everything
wrong, drunk on *clairin* by afternoon, from what I hear.
But if it slides, we shall fix it then, when the summer rains
stop new work.

From years of trial and error, judgment takes rude shape
in some inner chamber of the mind. The earth and its material is alive; it shows many faces, cooperates or rebels.
The artisan recognizes the personality of earth and stone,
just as men come to recognize friendly dogs or troublesome superiors. It is a perishable thing, this judgment; it
dies with a person, for it cannot be tamed in books, frozen
into codes and regulations. To some technologists its
purest form may even be obscene, as in the water witch,
the faith healer, the fortune teller. But take note,
technologists: it was *technē*, artifice, that brought forth the
ribbed vaults at Durham Cathedral; *logos*, systematic
knowledge, spawned the freeways of Los Angeles.

The road climbs to an altitude of four thousand feet,
terminating at the village of Kenscoff. There is a rum distillery there, featuring one of those tasting rooms that retail
preboxed selections of coconut- and banana-flavored rums
to tourists who arrive in diesel buses and Avis rental cars.
The tourists, finding nothing notably scenic or "Haitian"
in either the conventionally Floridian architecture of the
buildings or in the parking lot, gather in the tasting room
to sample the various rums, which, it is generally agreed,
are "unusual." After the sampling, their guides, those un-
trustworthy devils on whom they have so foolishly lav-
ished too much affection, delicately suggest the possibility
of a purchase. There is a quiet nervous moment. The wives
imagine serving at least the coconut variety while enter-

taining, hoping their husbands will explain the bizarre offering, that they "picked it up at Kenscoff, when we were in Haiti." Meanwhile they are on guard, the husbands; wretched children and women have been plucking at their sleeves for days now, and they have transferred their wallets to their side pockets. But their guide is saying, "Very good, from Haiti," and they think to themselves, what the hell, twelve dollars, old Jean Paul here, he can use the commission, whatever it is, with all those kids.

The simultaneous progress of these predictable transactions among the two dozen American visitors is perturbed by two German couples who talk loudly, taste everything, buy nothing, and finally, to everyone's relief, go for a walk in the hillside garden outside.

The departure of the Germans causes a perceptible elevation in the mood of the Americans; several begin chatting for the first time with their fellow travelers. The tonic effect extends into the small proprietor's office just off the main room, from which Monsieur Hechtman, who purchased the distillery upon emigrating to Haiti from Leipzig in 1938, emerges to join the tourists.

Hechtman is a middle-aged man who wears one of those expensive limp shirts of the kind featured in Bloomingdale's sportswear section. He has red curly hair and powerful wrists, and his damp skin is freckled with liver spots. At first I mistake him for a lonely tourist when he asks me how I like the overly sweet liqueur. It is more for the ladies, he says, smiling, before I find words to answer. "Ah, you live near San Francisco?" he says. "Very nice. Ernie's is one of my favorite restaurants." Hechtman touches my arm. Something in his manner says that there is, there should be, no fear of posing between us. He tells me of his coming to Haiti and buying the distillery from its ailing French owner. "He was putting too much wine into the making of rum. And your business?" "A consulting engineer; I have a small consulting firm," I say, wishing to establish our entrepreneurial brotherhood. "You have your own business, good," says Hechtman, speaking in a conspiratorial way that attracts but also alarms me. He is somehow making me want to please him, which weakens me. "There are some American consultants working in Haiti now," says Hechtman. "Not much, though. Matters are undecided, since Duvalier's death; it will be a few months before the situation is clear." Hechtman shrugs. The implication is that I am looking for business. It is true, of course; the idea that I may be losing my client, the chief engineer, flits through my mind. But I am annoyed; Hechtman is out of line in suggesting that he knows my need.

"It would be interesting to work here," I say. Interesting. Mark that. Not necessary for my survival.

"Cela va sans dire, that goes without saying." says Hechtman with an oily smile, and inviting me outside to see his garden. "You will find it interesting, I have built a path down to the stream, my engineering project."

There is something about Hechtman that I dislike. He has too quickly constructed a model of me, infuriatingly accurate, and now he deals with the model, not with me. Someone once told me that a good salesman can take inventory of personal memorabilia in a prospect's office—golf trophies, pictures of children, notes to call doctors—and build his sales pitch on that insight. Perhaps for that reason my office is as impersonal as a dental clinic.

We walk down the steep path, through bougainvillea, banana, and hibiscus. Hechtman speaks sharply to an old Haitian, a gardner; the man removes his hat and bows slightly. The path twists down the slope, behind Hechtman's house a cantilevered affair that looks as if it were lifted from the pages of *Good Housekeeping*.

Halfway down the hill, below the house, a flat pad has been notched into the slope, on which a crew of laborers are diffidently building a concrete swimming pool. Hechtman shows me the pipe that diverts water from the stream into the pool. A steep cliff about fifteen feet high has been carved into the punky red rock above the pool and below the house; two of the laborers are building a retaining wall of mortared cinder blocks to keep this soil and rock from sliding down into the pool.

Hechtman turns to me. "What do you think?" he asks.

So that's it! That's the point of all of this; he is looking for some free advice. Suddenly everything that has happened in the last few minutes fits together. I see the motive.

Looking at the wall, I slip into my analytical mode. Let us say that a massive wedge of soil and rock is on the verge of sliding, perhaps taking the house foundations with it. We have this wall of cinder block restraining it; if the wedge goes, then the wall must fall over. Now, what resistance does the wall provide? Could Hechtman and I and the three laborers, pulling on a rope attached to the wall, tip it over? It is only about a foot wide at its base; probably we could do it for a short length of wall—say two or three feet. I would have to work out a simple vector diagram to prove that, but I think we could. So the wall will offer some restraint, although not much. Now, back to the soil wedge. Could the two of us hold it back, if it started to slip? A thin wedge, yes, because the grains and particles of the weathered rock are interlocked, producing a sort of pseudo-cohesion, as long as the material is lightly confined, is not allowed to loosen. What about a deeper, more massive wedge? Well, there we should be O.K. because the natural slope is steep and did not slide of its own accord for thousands of years, so failure surfaces equal or flatter than the natural slope should be safe. What about medium-deep wedges? There I am stumped; my intuition fails me; I would have to work out the calculations, for which I would need good estimates of the angle of internal friction of the mass. I must suspend judgment on that one.

These thoughts pass through my head in about thirty seconds while I pretend to inspect the construction of the

wall itself. Meanwhile I notice that the vertical joints between the stone blocks of the wall have not been fully
mortared, so there are some open cracks in the wall.

A small drama can be created here. I stop to examine
these cracks, pick up a stick and jam it into one of them,
maintaining a perfectly blank expression. The stone ma-
sons have stopped working; all eyes, including Hechtman's, are on me. Hechtman, frowning, thinks I have
found in these cracks a flaw in the workmanship. Let him
think that; I've made my decision on that one. But back to
the overall stability of the wall, which still puzzles me.
Really, it is not much of a wall, too narrow at the base,
unreinforced with steel; it would never stand if the soil
and rock above it were loose or soft. But they have been
standing there now for perhaps a week or so, with no support at all. There are, of course, the rains to account for; a
medium-deep slide presumably will be more apt to go
when the slope is saturated with water, so the present
situation is not the acid test. Again, that same area of
uncertainty.

I shift my focus back to client relations. Time for some
meaningless talk that I can keep up while simultaneously
working on the problem.

I pick up a handful of soil. "This deep red color, it's iron
and manganese," I say to Hechtman as we begin to walk up
the trail. I ramble on about tropical laterization, how the
walls of Cambodia's Angkor Wat are made of manganese-
rich soils that were cut from the ground as soil blocks,
then allowed to harden irreversibly. Hechtman patiently
feigns interest. I am using the "irrelevant dissertation
ploy," in which the consultant disguises his uncertainty,
his anxiety at the prospect of an impending decision and
pronouncement. In this way one avoids "thinking out
loud," the mark of an amateur.

Halfway up the trail I stop and turn back to look at the
hillside and the wall. "What will be the consequences of a

failure?" I ask myself. I try to imagine the slope sliding, the slide propagating up the hill, spreading under the house, so that the whole house begins to slide. I think I can rule that one out; the rock is simply not that weak. How about a partial wall failure? That would surely create a mess; it would take perhaps half as much labor to rebuild the wall as it has to build it in the first place. Perhaps one-in-ten odds something like that would happen in the next few years. Nobody would get hurt because it probably would happen in a big storm when no one was using the pool. What would it take now to ensure that such a failure wouldn't happen? Doubling up the wall thickness, adding reinforcing. More excavation. Similar to the failure repair job, the same order of magnitude of cost and time. Hardly makes sense to spend a dollar today to prevent something that probably won't happen but which would cost a dollar five years from now if it did. O.K., that's enough; I'm converging on a solution. Put it back in the oven to cook.

The slope is heavily vegetated with banana, and I ask Hechtman about the trees, whether they bear edible fruit. Do they have several different types in Haiti? In the Far East, the ones with seeds were not very sweet but were said to be the most nutritious.

The conversation fades as we reach the top of the hill. Here at five thousand feet I am breathing harder. We are in the clouds. It is very humid, unpleasantly so, and now the swirling mist breaks and we are flooded with the sun's heat and brightness.

Something happens to me. I turn and speak to Hechtman. Well, you know Mr. Hechtman, I've not made a detailed study of your wall, there are various calculations and tests one can perform, although not usually on a small project of this size. Hechtman is nodding at this, accepting the conditional preamble. And I could find some flaws here, I go on, ways in which the wall wouldn't meet our codes in the United States. But aside from all that, I think what you want to know is whether this wall is reasonably safe and whether these fellows seem to know what they're doing. And my judgment on that score is that this is a good wall; it's the way I'd want it done if it were my wall, put it that way. Someone obviously knows what they're doing there, from experience. For example, I noticed that they left some of the vertical joints unmortared, those open cracks, and that's exactly the way it should be, to drain the water in the soil behind the wall, because water is the real enemy of walls, and yours would surely fail if they hadn't done that. So if I were you, I wouldn't worry at all about that wall.

Hechtman looked pleased, thanked me profusely, and invited my wife and me back into the tasting room over to the row of flavored rums.

"Please, each of you pick one, as my gift," he said.

Each of us picked a bottle, one cherry and one coconut. I couldn't find any plain rum, hating myself for being afraid to hurt Hechtman's feelings by asking for a substitute for the artificially flavored variety. The flavors were probably made in some chemical factory in Florida.

I mumbled thanks to Hechtman, exaggerating my gratitude in proportion to the degree that I did not feel it, for I had worked hard—Hechtman knew that—and now I felt humiliated by Hechtman's insulting payment.

Not until much later did I realize that Hechtman had unwittingly done me a great favor. I realized then that I had been living too long in the sterile world of analysis, in which theoretical consistency became an end in itself, divorced from the grit and sweat of reality. With Hechtman's wall, there was no opportunity for further studies; I had considered the available information, determined that it was sufficient for the case at hand, and in the end delivered a judgment, which I found good. Now, having come at last to Haiti, I understood what Professor Williams had been trying to show us in that gloomy Cambridge winter fifteen years before. Having failed to see the vision, I had, like Daedalus, lived for many years in a stone tower of logic and analysis. Now at last, in this steaming island in the sky, I had spread my wings and learned to fly.

I thought when it all began that it was going to be like any other business or engineering problem, something to tackle with logic and flow diagrams and Bayes theorem. And so I was surprised to find myself sitting in that chair, a lawyer circling my flank like an unfriendly dog on a dark street, and the flat look of the judge telling me my credibility had just slipped a notch; surprised that it was more like saying good-bye to a brave child in a hospital or the first week in boot camp. It just snaps through the polished surfaces of one's professional armor with the ease of an electrical charge, sets the most private of the smooth muscles to dancing. There is no refuge; at three o'clock in the morning, fermented bits and pieces of the day hiss from the recesses of your mind like gases from buried geologic strata.

They came after me when I was in bed with the flu.

The long arm of the law, they call it. To me it seemed an elephant's proboscis, the exploratory organ of a lawsuit, its delicate pink tip, mucoid and hairy, searching for an unshelled peanut. Backed by an irreversible force, feeding an innocent but insatiable appetite, it glided expectant and trembling into my office, and touched my secretary. "Out with the flu? Oh that's too bad. I suppose he's home, then?"

It was a hot afternoon in August, and I was motionless, impressing with my back a steamy dampness onto a white sheet. The fever was a brass wafer on my tongue, my skin as fine tuned as a seismometer. I had two symmetrical but distinct headaches, one behind each eye. I confess to taking secret pleasure in the flu: it is one of the few safe and legitimate refuges of middle age, a bright temple within a grove sickly sweet with blossoms of orange and almond, grounds of raked sand forbidden to salesmen, business associates, and children; friendly spirits visit me there like schools of tropical fish. It was four o'clock in the afternoon. Only the most exquisite sounds came to the room where I, the patient, lay; birds twittering in the happily spattering lawn sprinkler; the bland conversation of the mailman; a butterfly in the garden of women and children; Elton John, faintly, from my daughter Kathleen's room. My wife had gone to the marketplace. She would return with offerings, fruit peeled and cut in a china bowl, clean sheets, oatmeal, a fresh *National Enquirer*. I lack the nerve, at my age, to ask for a Superman comic book.

The doorbell rang, and Kathleen thumped down the stairs, yelling, "I'll get it." Paper boy collecting for the *Times*, I supposed. Or one of my neighbors petitioning against the affront of nuclear power—probably the bearded one two doors down with the dog that howls all night. I'd gladly trade both him and his dog for a few extra rems of radiation. Or a child, speaking a language I barely understand, selling inedible candy wrapped in magenta foil. Kathleen will handle it, whatever it is beating at my door.

There were mumbled words, and Kathleen came to my room. "There's a man who says he has to see you. He says it's important," she said.

"Tell him I'm sick in bed, asleep; ask him what he wants," I told her. Not a clear reply. Am I not leaving the door open just a crack here? Perhaps I am finding the solitude of the temple a little boring, succumbing to a dangerous curiosity about the world. Or am I seduced by "it's important"? Meaning I am important?

Kathleen returned to the front door. There was more subdued conversation. A moment later she came back, and there was a man with her. Slight, balding, with glasses, clean shaven, a short-sleeved shirt open at the collar, he stood nervously at the threshold of my bedroom. He looked like an electrical engineer. He held a folded piece of paper.

"Mr. Meehan?"

"Yes."

"This is a subpoena."

The man inspected his conscience, which he evidently kept beneath his fingernails. "Hey, I'm sorry about this; this isn't my usual job, delivering these things; I'm a detective. But the regular guy was out today."

Hard times have come to the detective business. What with the no-fault divorce laws in California, there's no romance in it any more; a man can't earn a living following wayward husbands into seedy hotels. Sam Spade reduced to delivering subpoenas to the sickroom. Decay of a proud profession.

I accepted the paper, feeling sorry for him. For a moment, I wondered whether I should have offered a tip.

The construction industry had slumped badly in the autumn of 1969, a few months after my partners and I had started our little consulting engineering firm. In those days, even the better established consultants were scratching for work. Like many other engineers going into business for themselves, we had underestimated the persistence and flair that it took to bring in new clients, at least ones who pay their bills. My father-in-law, a successful surgeon and banker, had loaned me $5,000 to start up this business, and the money was beginning to run out. It was a rainy gray autumn, and I was spending my days in San Francisco, visiting prospective clients who had no need for our services but who welcomed me politely in their steamy offices and listened patiently to my nervous pitch. I had one good pair of shoes; they grew sodden and heavy from the wet sidewalks, and I developed a bad case of athlete's foot. My wife devotedly ironed a clean button-down shirt and pressed the crease back into the rain-soaked, shapeless lower part of my trousers every day. She never asked about "new jobs" unless I brought that matter up. At the office, we were making ends meet, but just barely. A new project, especially when it was a new client, was cause for celebration.

One of my partners, Irwin Sprague, was active in Republican politics and through those connections had been introduced to the building director of the San Ignacio School District. The town of San Ignacio occupies a valley of the same name, a wrinkle in the California coast ranges south of San Francisco. Until the mid 1960s, the floor of this valley was carpeted by matted bunch grass and desiccated cow pies. Now it contains an inland sea of pastel tract houses, warping and fading in the sun. The district had just acquired a site for a new high school, and the director contacted us about performing the standard soils investigation of the site and advising their structural engineer and architect on matters of foundation design. We were delighted and worked up a program to perform the necessary borings, soil tests, and analyses. We would charge them in accordance with our standard hourly fee schedule, we agreed—twenty-five dollars an hour for our time, plus expenses, with a guarantee that we would not exceed three thousand dollars.

Looking back on it, we were offering a real bargain, even for those days. Jobs like this were usually assigned to junior staff people in firms like ours, but we didn't have any of those, so Irwin and I, with twenty years of experience between us, handled the field and lab work and report writing ourselves. We were eager to do a good job—perhaps overeager, it occurs to me now. There is such a thing as trying too hard, and it affects one's judgment. It takes experience, and perhaps age too, to understand that professional skill is a peculiar alloy of strict standards and callous indifference.

We knew we had trouble the first day on the project when Irwin returned from drilling holes at the site. "It's a fat clay," he said, placing a lump of malevolent black adobe on my desk. I ran some tests on the soil, which confirmed that it was a fat clay, the kind that shrinks when it dries and swells when it's wet. That means trouble for building foundations in California, with its wet and dry seasons.

We sent in our report and then did not hear much about the high school for several months. One day the following spring, someone told me that the architect and structural engineer were preparing final plans and specifications. This got me thinking about the job again, especially about the two alternative solutions I had suggested. One of these was to dig out the fat clay beneath the school and replace it with gravel. The other was to mix exactly the right amount of water with the clay so that it would neither shrink nor swell.

Then I had some second thoughts.

True, the special moisture-conditioning solution would save the owner some money (we engineers are steeped in an aesthetic imperative that demands functional adequacy at lowest cost). But it also assumed that the contractor would know what he was doing and that his work would be closely supervised. How did I know whether the district would get a contractor who had any experience in this type of work? Would they have someone competent supervising the work in the field? No one had ever bothered to discuss the plans and specs with me; could I count on the designers having incorporated all of the special provisions that I had recommended for using the on-site soils? Obviously I could not assure myself on any of these points; they were all out of control. Maybe, all things considered, it would be better to stick to the simpler solution that called for replacing the soils at the site.

I put a call through to Frank Schultz, the structural engineer, telling him of my concerns, that I was inclined to go for the soil-replacement solution, the safer one. Schultz disagreed. He was an old-timer, and he liked to run his jobs. Once he had made up his mind on a point, Schultz had an effective, rhetorical, direct-mail manner of presenting his case. He said that the district had just about exceeded the budget on the job already, and every penny counted. He had a lot of precautions built into his design, extra heavy reinforcing in the concrete and special protective barriers to keep the water out of the foundations. He had designed hundreds of buildings on soils just like this, or worse, and never had problems before. He wasn't worried about it. Schultz was a well-known designer, with a lot of experience, including many structures in areas where I knew the soil conditions were even worse than at our site. O.K., I told him; I agreed that we should go ahead with the cheaper alternative solution. We chatted for a few more minutes about design details, then rang off.

That was the last I thought about the high school until late that summer, when Irwin came by my office one afternoon and told me he had got a call from his friend, the building director at the district. They were starting construction in a few days and would be calling on us from time to time to perform soil tests to check the moisture conditioning operations. They had their own man on the job to supervise the work from day to day, so all they needed was a few tests, at his direction. Irwin had discussed the frequency of testing with them a bit and concluded that one of our technicians could perform the work in about thirty hours' time. He suggested they budget six hundred dollars for our testing services on the job.

Matters went downhill from there. The district had reluctantly awarded the job to a local contractor, Mullins, an affable charmer with a spotty record of past performance. The district's man on the job was a beery construction stiff hired from some other public agency.

I assigned a recent graduate with only a few months'

experience as our on-call inspector. The site work started, then stopped for a few months because of a strike, then had to be partially redone, then started up again. For one long wet winter, the uncompleted work was lashed by rainstorms, with no effort being made to protect the sensitive foundation soils. Half-completed precautionary measures that I had specified, drains and moisture barriers, perversely became the conduits and baffles through which storm waters gushed into the foundations. The contractor's men appeared to be moonlighters with little experience in the delicate concrete work required for the job. Schultz, the structural engineer, raged over unacceptable workmanship from time to time, and some of the concrete work was ripped out, redone, and then ripped out again. Anything that could have gone wrong did. In the end, the district threw Mullins off the job and demanded that his bonding company finish it in time for the school year. They did, but it came out looking patched and second rate. A month before the dedication, a huge crack appeared in the floor at the main entrance. All of us who had been involved with the project were ashamed of it.

Three years later.

The date of my pretrial deposition loomed before me, piquant and dreadful, like a final exam in organic chemistry. Irwin and I had the idea that we should hire a lawyer. Our consulting business had enjoyed a modest success since that time four years before, when we had worked on the San Ignacio High School project. The project had turned out a mess, with cracked slabs and warped walls. The school district was blaming the contractor, Mullins. Mullins was blaming the district's consulting engineers, Frank Schultz on structures and us on soils. Now the issue had become a $150,000 lawsuit, in which we were named. We had no malpractice insurance, so we were on our own, had to pay for our own defense. I had never been involved in a lawsuit before. They said that an experienced trial lawyer charged fifty dollars an hour. And up.

I had a lawyer friend, a bright, droll Harvard man, Henry Worthington. Henry had just finished making some mid-life adjustments—getting a divorce, going on the wagon, and quitting his job with a big city firm. Putting myself in his hands would probably cost a few hundred, I figured, but it would be worth it if he could make it all go away.

Henry, the new Henry, had a regular schedule on Monday, Wednesday, and Friday that began with meditation and granola at 5:30 in the morning, followed by a session of vegetable gardening using the French intensive method. Then he spent an hour reloading shotgun shells, in preparation for his afternoon skeet shooting. Henry had elevated skeet shooting to the level of a spiritual activity, the transubstantiation of clay discs to puffs of black smoke. I met him at the range one windy afternoon, shot seven out of twenty-five, then, unenlightened but with my ears ringing, asked him if he would review my file and come with me to this deposition.

Henry was neatly packing empty shells into a box, and he paused and looked at me. He is one of those people who can pause in a conversation for thirty seconds or so, create with perfect calmness a frightening void during which you fear that your heart will stop and you feel an urgency to explain or confess something. No, he could not handle the matter, he finally told me without explanation. Could he recommend someone else, I asked. Henry thought about that for what seemed another geologic age, then began to critique my shooting style. I was responding to the wrong kinesthetic cues. "Maladapted sternomastoid response is where the major problems begin," he began to explain.

"I know, Henry," I said, exasperated. "I think about all those things too, but I'd really appreciate it if you could just give me a name."

"Try Duncan Macaulay. Duncan's a superb litigator, just

superb. A lone operator. You don't find many of those around any more."

I made an appointment to see Macaulay, and a few days later Irwin and I went to his office, a small building he shared with some other lawyers, tucked inconspicuously into a leafy residential street. No stable of young legal wizards to keep busy at forty dollars an hour, I thought. A good sign.

Macaulay proved to be a tweedy bearded chap, fiftyish, with a cagey opening manner that immediately struck me as potentially effective against our adversaries. He made no effort to convince us that he knew anything about our business or the various technical matters relating to the case, which I thought reflected favorably on him. In my experience, playing up credentials in or knowledge of the client's business is always a sign of weakness.

I explained our background on the project, that we were not too concerned about losing the case because there was no question of any negligence on our part, but that we thought we might have some exposure here, that we didn't carry any liability insurance against this kind of thing, that the insurance wasn't even available anymore.

Macaulay shook his head in the socially sympathetic manner of one unmoved by others' troubles. "You guys really land up carrying an unfair burden in these situations," he said.

This annoyed me. I didn't want professional sympathy. We were interested in making the lawsuit go away, or failing that, going for the jugular veins of the contractor, the district, whoever might be after us. But Macaulay looked so perfectly insincere in this remark that I didn't hold it against him. We agreed to retain him to work on the case, and he marked his calendar for the day of my deposition.

A couple of days later, Macaulay gave me his first piece of advice, which proved, in the end, to be bad. Or perhaps the mistake was in my asking his advice on the matter in the first place, not making my own decision.

For the past year or so, I had kept the fat manila project file for the high school job at home. Not that there was anything incriminating in the disorderly pile of pink telephone messages, photocopied letters, fold-softened ozalids, and mud-splattered calculation sheets contained in it. But I had an anxious vision of a mob of detectives bursting into my office and cleaning out my files, then something turning up in them later that would prove embarrassing to me. I make mistakes and overreact to situations and use bad language just like everyone else. It might well be that the law says that some lawyer has the right of access to all that, but that's someone else's law, not mine.

I read through the file a few times, then threw out a couple of items that made us look more foolish than criminal. There remained pages and pages of field notes and calculations, and I almost threw those out too. But the subpoena instructed me to bring all my project files and notes to the deposition. I called Macaulay and asked him what to do.

"You don't have anything to hide, do you?" he asked me.

I wasn't aware of anything. So in the end, I followed his advice and brought the files (which the other lawyers photocopied) even though it ran against my own intuition. Lawyers are skilled at explaining the rules. But there are certain decisions that you had better make yourself, at midnight, alone in your office. And the point to keep in mind at that time, and at all other times, is the more you give the opposition, the more they have to work with, the bigger the ore body they have to mine for the rich veins that occur in every barely perceptible crack in your professional performance.

The day of my deposition arrived in the middle of a late August heat wave. I woke up at seven o'clock that morning to find the heat seeping through my bedroom window, displacing the glade of moist air in which the filaments of my mind had finally disentangled in the gray hour before sunrise. By the time I had reached the San Ignacio Valley, an hour's drive from my home, the morning had grown sullen and hot. Mists that during the night stroked the feverish hills like a mother's cool hand had now drawn away, and the air itself felt thin and burned by the orange sun.

I found the attorney's office, where my deposition was to be taken, in an abandoned orchard on the edge of the town. The building was a new California missionary-style stucco and chicken wire tax write-off with an imitation tile roof, partially screened by parched olive trees and surrounded by a sticky petroliferous parking lot.

Inside a receptionist with carmine lipstick and a rococo tower of hair told me that it would be a while and suggested I join a large plastic plant that seemed to have blundered into the waiting room. I flipped through a copy of *Audubon* magazine. An hour passed. An hour and a half. The day was emptying itself of otherwise paying work. I could feel whatever organs of rage exist in the region of the solar plexus inflating. Aware of the hazards of anger, I ransacked the files of my mind for consolation. I told myself jokes. I visualized an advisory panel consisting of William James, Norman Vincent Peale, Marcus Aurelius, and Doris Day.

An hour later the session recessed for lunch.

The morning's victim had been Frank Schultz, the structural engineer on the project. Frank and his lawyer (actually his insurance company's lawyer), my lawyer Macaulay, who had sat in on Frank's deposition, and I walked across the street to a Sambo's restaurant for lunch.

We ordered our burgers, and the insurance company lawyer, Don Smith, up from Los Angeles for the day, coached Frank on his morning performance. Smith wore a polyester suit that contained, under some tension, the bulk of an ex-athlete gone soft. I tried to read the inscription on a gold and ruby college ring he wore on one fleshy finger of his big left hand. I was curious because I couldn't determine to my own satisfaction whether his diffident commentary on the morning's proceedings were the gropings of a sluggish intellect or the faint output of the teaspoonful of brain matter that he was willing to assign to the conversation, holding the rest perhaps for a current love affair or a faltering investment.

Frank was vibrant with emotion and was not listening very closely to his lawyer's advice. He was in fact repeating, to my benefit, some errors he had made during the morning's session. "When I said that that contractor Mullins was incompetent, I meant it in the full sense of the word," he blurted. Contractor incompetence was how Frank saw the whole issue. It was Frank who had advised the district to discharge Mullins from the job.

Smith asked him if he really had authority to direct the contractor's work in the field.

"This was my job, and I had full authority to tell him what to do," Frank said angrily, as if challenged by his lawyer.

Smith frowned and said that he didn't want to tell Frank what to say on the stand or advise him to speak untruthfully, but couldn't he soften that a bit?

It was apparent to me that Frank was blindly digging his own grave deeper, by claiming to hold responsibilities on the job that he actually did not have at all. Under fire, Frank was taking hits and going down with pride, the captain of the ship. What he should have been saying was that he was only the galley cook, that on the morning the battle was lost he dished out the chipped beef on toast just the same as on any other morning. Perhaps we engineers tend to make that mistake, put too much of ourselves into our work, I reflected. Maybe Smith had the right idea. But Smith had now done what he could. His eyes glazed at

Frank's further protestations, his mind seemed to depart to Los Angeles, to whatever other arena was preoccupying him.

Smith's abandonment of the issue left the field open, and there was a silence that we filled with baconburgers tasting of substances half-prepared in distant cities. I decided that I would see what I could do to deflect Frank's self-destructive momentum.

"Frank," I said, "You've been in this business a lot longer than I have, but I'd like to say that in the last thirteen years or so I've worked as project engineer on quite a few projects, and I've never on one of them considered that I had direct authority to tell the contractor or his men what to do in the field. My role on the job, as I see it, is to advise the owner if something is going wrong; then he takes action if he chooses to."

Frank had relaxed and grown reflective during this speech and had evidently begun to listen; both lawyers were nodding vigorously in agreement.

"Right, Frank," said Smith. "Aren't you, as a consultant, really just an adviser?"

Frank thought for a moment, then began to nod. "Well, I guess you're right," he said. "I can't run everything."

I thought that even if, in my own deposition, I set our cause backward with my own mistakes, I had made one small contribution; my day would not be a total waste.

But in fact my deposition seemed to go well. I had never been deposed before and was surprised at how easy it was. In fact, it was fun in some ways, matching wits with the attorneys, always trying to say the thing that peels away that papery veneer of strutting indignation, exposing the underlying venal whine. I remember thinking that I would have to be careful during the trial, if it ever came to that, to avoid enjoying this sport; a man at pleasure is a man off his guard. But when I read my deposition a few weeks later, it seemed clean. It was typed with IBM executive boldface on crinkly white rag bond, flatteringly bound in a plastic cover. The day had cost us about a thousand dollars. Legal larceny is carried out in gentlemanly style; the victim's vanity is enhanced even as he is being robbed.

Summer passed, and in the bright crisp beginnings of fall my attention moved to other things. From time to time faint vibrations reached me as each new specimen of witness hit the web. The sullen ready-mix contractor. A retired engineer who reportedly held advanced degrees in concrete and offered expert witness services at a discount. It swayed and shuddered when struck with the district's three hundred pound field superintendent, but his struggles were brief and his submission to all interrogators so complete that Macaulay told me with satisfaction that he would not be of any use to anyone.

There was sporadic but diffident talk of settlement. "If this thing goes to trial," Macaulay told me one afternoon when he dropped by the office, "it could end up costing you five or six thousand dollars. You really shouldn't be carrying any of the blame here, but just from a practical standpoint, we're thinking about getting up a pot to get this contractor off our backs." I told him he could consider himself authorized to bargain up to three thousand. No, he couldn't see that, Macaulay said; fifteen hundred seemed like a reasonable maximum to him, for our share. But Mullins was claiming $150,000 from the district, and Macaulay told me a few days later that he had not shown any interest in the meager six thousand that our reluctant alliance—the district, Schultz's insurance company, and my firm—had put together. It looked as if the case was going to trial.

There was one other thing, Macaulay said. The district had decided to file a cross-complaint against Schultz and us. That meant that if the contractor won his case against the district, the district, our client, would go after us.

The trial began one Monday morning in February at Oakland County Courtroom 5C, a solemn, walnut-paneled room that reminded me of the Boston Harvard Club, which I used to visit with my father when I was a boy. I wore my best tweed jacket and regimental necktie.

I found Frank Schultz sitting in the visitors' gallery. Both of us greeted Mullins with funereal nods when he entered. Mullins was chewing a toothpick; he jammed his bulk into a seat on the other side of the aisle, returning our greeting with an affable grin.

The three of us were joined by one of those bleary old men who find courtroom drama preferable to television in some forlorn residential hotel lobby. That made four spectators. Not exactly the trial of the century. But that morning it seemed important enough to me; I had already poured out $5,000 in legal fees, which had done little to stem the unmistakable and ominous concentration of attention on my role as project foundation engineer. Soil and foundation engineering is supposed to be a rational business; in reality it is a soft and vulnerable nexus of structure and earth, where design objectives and the whims of nature, the responsibility of consultants, and the risks of ownership all blur together.

The lawyers sat at a long table before us. At one end was the district's lawyer, Tony DeLuca. Tony had entered a cross-complaint against my firm, putting me in the difficult tactical position of having to side with the district against the contractor, and at the same time defend myself against my own client, the district, by pointing out their contribution to the failures. "I really hated to do that," Tony told me one day, adding that the district might sue him if he had not targeted all possible pockets. Lawyers never mean any harm, of course. Nothing personal in taking away your eight-year-old Dodge Dart and the kids' college fund. Tony drove a new Alfa Romeo and wore a fresh, crisp suit every day.

Schultz's insurance company had a lawyer on the case, and he sat next to Tony, and next to him was someone representing the contractor's bonding company. Then Mullins's lawyer, a nervous, bearded villain named Lyons. With Lyons was his young apprentice, Vincent Moore, a quiet and conservatively dressed recent law graduate with a direct gaze that by the end of the trial would pierce the privacy of my dreams. And my lawyer, Duncan Macaulay, pulling at his beard and fumbling ursinely through his papers. Six lawyers. Three grand a day.

Lyons opened the trial with his kick-off statement of Mullins's position, an intense finger-thrusting speech packed with imitation outrage. The other lawyers followed, each speaking with a kind of impassioned rhetoric that was supposed to convey conviction in their client's cause. To an observer peering through the small glass window of the courtroom door, there was a ghostly semblance of brilliance, conviction, and wit, but to those of us in the courtroom, including Judge Bradford, a florid, balding man, fiftyish, with hard intelligent eyes and a dry manner of asking blunt questions, the arguments seemed unplanned, adventitious, and rhetorically untidy. In his opener, my lawyer Macaulay made a speech about how His Honor was going to have to be cautious lest he become confused by the complex technical issues involved in the case. He emphasized the critical importance of technical terms, some ill-chosen and irrelevant examples of which he then proceeded to define in a sort of sonorous and highly erroneous glossary. About a third of the way through his benumbing discourse, Judge Bradford began to fall asleep.

I was horrified, and later embarrassed, when Macaulay asked me on the way home whether I thought his opening statement had been effective. "I don't know, Duncan; you know what you're doing here," I said. "I've never been to a trial before. I don't have a television, so I don't even know what's supposed to happen in trials."

Mullins was the first witness. Lyons put him on the stand the next morning, Tuesday. By Thursday he was looking pale and addled, and Friday he called in sick. They finished with him the following Monday. "At this rate, this trial might go on for a month," Tony DeLuca said to me Monday afternoon. I was astonished, having once been found guilty of a speeding charge in thirty seconds. I had reckoned on two or three days at the outside. But I did not have time to brood about my mounting legal costs. Lyons announced that he was putting me on the stand the next morning.

There are, it should be understood by the inexperienced, certain dangerous delights in litigation, pleasures not unlike those that can be obtained at less expense in a game of seven card stud. It should have been clear by now that there was no way for me to win the case. Winning is beside the point when you are paying legal fees. I had already lost; the question to be resolved in the trial being only how badly I had lost. The drama of the trial obscures the simple fact that the real contest is not between the various litigants but rather between the litigants and the attorneys and that the attorneys always win. And yet there is a certain moral passion and euphoric stimulation in a courtroom that makes the entire proceeding a fascinating adult entertainment, a sort of puppet show directed by the legal profession, in which the litigants are offered fantastical rewards (the judge, in a ringing voice, "The court finds this defendant NOT GUILTY," cheers, tears, joyous laughter). And so it was that I went to bed that evening before my first day of testimony pleasantly intoxicated with a sort of precombat, psychic tumescence.

The next morning I learned to my surprise that I was to be examined as an unfriendly witness by Vincent Moore, Lyons's young apprentice, rather than by Lyons himself. It began well, I felt an easy confidence, and Judge Bradford seemed to be placing a high degree of credibility on my testimony. From time to time, to the evident annoyance of both lawyers, he would break in and ask me a few questions himself, my opinion on this or that. Moore got bogged down and confused several times and almost had his entire line of inquiry, which was obviously aimed at discrediting my work on the project, cut off by the judge when he failed to establish its relevance to his case. "This is his first trial," Macaulay remarked during lunch break. "He's still wet behind the ears."

But Moore, with a little help from Judge Bradford (whom I suspected of musing paternally on his first trial), recovered quickly. Moore wore a dark suit, a white shirt with that chiaroscuric gleam that new shirts have the first time back from the laundry. Earnest and spare in his manner, he seemed to have stepped not from the streets of Oakland but from the glazed shadows of a fine old oil painting. When he slipped in the course of his examination or his way was suddenly blocked by a sustained objection, I could follow and empathize with that flicker of uncertainty, the brief and dignified silence in which he walked back, head bowed in thought, the delicate rearrangement of papers, then the smooth recovery. "Mr. Meehan, is it not true that . . ." he would begin on his new tack in a clear voice that sprang, it seemed, from no other source than the purest desire to get to the truth of the matter. I could not help but admire his style, as he turned back with that question, one foot advanced before the other, chin raised, fingertips on the table, fixing me with that solemn Rembrandtesque gaze. Lyons, slight and black bearded, lurked in the background with the nervous shiftiness of a Shakespearean villain. He would have been easier for me; he had judged well in giving the examination to Moore.

But yet other reasons for Moore's assignment surfaced

later that afternoon when he came to the matter of our field soil test calculations. Evidently Moore had an undergraduate background in mathematics, and he had exhaustively checked every one of the hundreds of calculations that my inspector had made in the course of our construction soil testing. And he had, it appeared, found errors. I managed to stave off the impact of this by remaining calm and seemingly indifferent to these disconcerting exposures, but I was worried, and when court recessed for the afternoon, I called Roger, who had been the field technician on the project, and told him to meet me at the office that evening. It became clear to me on further examination that there were quite a few mistakes in the arithmetic, and I told Roger to run the whole batch of calculations through a computer. By midnight we had the answers, and I prepared an exhibit comparing all of the computer results, printed in that mechanical bank-statement style that looks so convincing to the layperson, with the original slide-rule field calculations. True, there were differences in the results, but the comparison did not look nearly so bad when we looked at all of the tests instead of just the ones with mistakes.

The next morning, Moore resumed his examination of every inaccuracy he could find in our mud-splattered and partially illegible field calculations. ("And do you consider, Mr. Meehan," with arched brows and mock astonishment, "THAT particular error to be acceptable work in your company, or the standard of your industry?") But after an hour or so of this, Judge Bradford was looking sleepy again, and the time seemed right for me to suggest we would save some time by examining my computer review of all the calculations, not just the ones selected by Moore. Moore hesitated, glanced at the judge, who seemed interested in my suggestion, and reluctantly agreed. My computer display blunted the impact of Moore's attack, and for the first time in hours I felt as if I had regained some control over the situation.

Moore moved on to other areas, showing how from time to time my inspector Roger and I had given field instructions and approvals that differed from what was required by the written specifications and how in one case I had rather arbitrarily "corrected" certain test results. I kept explaining for the judge's benefit the overriding importance of field judgment and experience, but by mid-afternoon I felt I was losing ground. Now the various parts of my mind, which the previous day had dodged and whirled with regimented precision, began to come apart, like a troop of boy scouts at the end of a long hike. While its leading elements forged ahead, Moore was on its flanks and rear, working on the weak and straggling member. But some very tired scoutmaster in me continued to patrol the line and pull back the stragglers, all the time maintaining a feigned scoffing attitude toward the attack.

Then suddenly, right at the end of the afternoon, Moore shifted his line of questioning and asked me whether any of our tests showed evidence of soil movement. After two days defending myself on the stand, I was primed for denial.

"I don't have any indication that any movement of soils occurred on the site at any time," the record shows I said.

Judge Bradford looked at me carefully and made some notes.

Afterward Macaulay and I talked about the day on the way home. I had the feeling that I had overstated my position, but by that time I couldn't remember what I had said. Perhaps we could straighten it out on cross-examination. "Up to today, you were clean," said Macaulay. "Today they grazed you."

That week of my testimony I spent a lot of time with Macaulay. In the morning I would meet him at his office. There Macaulay, backstage, would rush about talking in a distracted way, stuffing papers into a briefcase, shuffling through the pile of mail that was accumulating on his desk. Meanwhile I would foolishly dose myself with plastic cupfuls of bitter machine-made coffee, for which I would pay later on the stand with a full bladder and an annoying hum in the wiring of my nervous system. As usual I would resolve to learn to chew gum instead of drinking coffee.

Then in Macaulay's gray, coffin-like Mercedes, the two of us would make the one-hour drive to Oakland. Crossing the flat waters of San Francisco Bay on the Dumbarton Bridge, we would skirt Newark, its new subdivisions bordered with bewildered and homesick eucalyptus trees, then enter the river of northbound traffic flowing lazily through Union City, slowing to a stagnant pool at Hayward, then breaking into a grimy ripple at San Leandro. Oakland lay beneath a cloudless sky streaked with amber turpentine wash, stinking of inner tubes and mercaptins. I watched the litter-strewn streets, with their steel-caged liquor stores and abandoned stucco gasoline stations, flip past below the elevated freeway.

At those times Macaulay and I would discuss the case, and sometimes I would probe him for cost estimates. It had been my custom from the start to maintain a record of my best projection of the total cost of the suit, it being a theory of mine that a best-guess cost estimate and estimated coefficient of variation are necessary at all times when continuing policy decisions are required. Of course, as the trial went on, I had to increase my estimate continually. Macaulay alone was costing $500 per day. And by the third day of my testimony, I was thinking in terms of a six-week trial. My working cost estimate, including legal fees, my own time, and an allowance for a most probable settlement cost, had gone from $5,000 before the trial, to $15,000 at the beginning of the trial. Now, in the third day of the trial I was figuring $17,000 in legal costs, $5,000 in lost work time, and $15,000 in settlement costs, realizing that in the worst case the last figure could be much worse.

In this way, my best-guess estimate had gone from $5,000 to $37,000. It was an ominous trend, what would appear to be a deterioration of our position, or to put it another way, evidence of a very bad initial estimate, a failure to be realistic about the situation in the first place. With some annoyance I recalled Macaulay's early advice that we should offer no more than $1,500 in settlement. Couldn't he have foreseen that there was a significant probability of this $37,000 outcome?

The next day, Thursday, was my third on the stand. Moore, having got out of me late in the afternoon the denial that any of the tests had indicated soil movement, decided to quit while he was ahead. So the next morning I was examined briefly by the other attorneys, then cross-examined by Macaulay. He made a number of false starts, was blocked successfully and to his great annoyance several times by Moore's clever objections, then finally had me run through a demonstration of field test techniques that seemed to me irrelevant and again put the judge to sleep.

Earlier I had told Macaulay that I wanted to restate my position on the soil tests because I felt I had made an incorrect statement, and during the afternoon session he got around to asking me about what the tests really showed. Had any of them indicated soil swell? Yes, some of them had indicated that, I answered.

Judge Bradford came alive. "That's not what you said yesterday. That's not what my notes show." He gave me a long look; his eyes were like two small aluminum discs.

I couldn't remember exactly what I had said. "Your honor, I hope I didn't say that; my position is that the tests, some of them, showed an increase in soil water content, and that is one indication of swell."

Bradford said he wanted to review my previous day's

testimony on that point. The court reporter fumbled with the stacks of transcript tape. "We'll look at it next session," Bradford said. "Court dismissed."

The next day, Friday, was a court holiday. That left me until Monday to find out whether I was going to be charged with perjury.

The next day I went into the office, presented the grim outlook to my partners over lunch, then took the afternoon off. Friday evening I went to bed early and slept until Sunday morning, waking up for only a few hours on Saturday. It was that kind of tempestuous dream-filled sleep from which you arise drained and exhausted. I remember one dream: Judge Bradford changed into a woman with hard eyes and bloodless lips. I recognized her; she was an East German border guard who years before had held me up at the border because of "irregularities" in my passport.

Monday morning I was on the stand at 10 A.M. I had my files with me and my journal, in which I wrote notes whenever I had a break. The calming effect of doing two things at once. Moore was going to reexamine me.

10 A.M. Sitting on the stand, waiting for Judge Bradford to come into the courtroom. Before me the five lawyers, ranged like an avenging army. Heart rate a bit higher than normal, but hands steady, not excessively nervous. Court reporter enters, everyone stands. The judge will be a minute or two, he's on the phone. So far I've been preserving a cheerful and calm demeanor. Can Moore push me to the point of cracking? I don't know. Feeling pretty alert and rested this morning. A buzz. Judge enters.

During the eleven o'clock break Moore asked me if I were writing my memoirs. "Yes, I'm writing about you," I said, hoping to unnerve him.

"(Unintelligible) sold his before he wrote them," Moore said to Mullins.

"Who?" I ask, thinking he said Nixon.

"Kissinger."

1:45 P.M. Just finished putting sketches on the board, illus-
trating various types of slab misalignment. The drawing
looks good, and I am feeling confident regarding my testimony (the contradiction in my testimony of last week has
been handled and accepted as a simple "clarification").
Macaulay and the judge are now conferring on some obscure point of law.

For the last hour of my testimony, Moore worked on me
in his calm manner, with his partner Lyons occasionally
leaping from his chair, a sort of Roderigo, advancing at me
and my diagrams, thrusting and parrying with a wooden
pointer. I had my own pointer, an aluminum one, and
there were moments when a still photograph would have
caught us in a duel.

2:45 P.M. Finally finished my testimony. No disaster today.
Managed to fend off Moore by evasion, frustrating him.
Playing stupid. This is O.K.; isn't it true that the examining
attorney extracts testimony by appealing to the professional's vain need to know everything? By asking for help
and cooperation? What emotions interfere with my func-
tioning perfectly as an expert witness? For one, the fear of
being exposed as inadequate, the guilt over mistakes, exposure of the fact that "my" project didn't work out adequately. Remember your advice to Schultz!

My testimony was finished, but of course the trial had
just begun. There were Schultz, other parties to the suit,
various experts on concrete. A month later, the trial was
still going strong, five days a week, with no end in sight.
Judge Bradford hinted that the six lawyers were dragging
it out and pressed for a negotiated settlement.

The problem, everyone agreed, was that the district's
board of directors was unwilling to negotiate a settlement.
The rest of us long ago had replaced moral righteousness
with cynical pragmatism, but the district consisted of
elected people, and elected people do not come quickly to
this point of view.

Hearing that Tony DeLuca, the district's lawyer, was

meeting with the district's board to brief them on the trial, I decided a bold move was in order. Never mind the goddamn lawyers. I wrote a letter appealing to the district board members directly, telling them that we would continue to support their case vigorously against the contractor but only if they would drop their cross-complaint against us. Otherwise, I hinted darkly, we were in the untenable position of being threatened by them, our allies. Inevitably it would be in our own interest to expose the district's mistakes that contributed to the problem, which would only help the contractor. Before writing the letter, I spent an hour reading some of Thucydides' accounts of successful oratorical appeals made by various envoys during the Peloponnesian War, to get the right flavor.

My partner, who knew a couple of board members, brought my letter to the meeting and tried to sit in on the conference, but the staff closed the session, although they did agree to distribute copies of our letter to the board members. Irwin waited for two hours while the meeting took place. Afterward DeLuca told him the board had voted three for, two against, to make a settlement offer of $50,000, provided it was split three ways—between district, structural engineer, and ourselves. That was $17,000 for us. Irwin and I agreed that we should tell them to go to hell.

The next evening I called Macaulay, told him that we wanted to drop out, stop supporting the district's case at our expense.

"You mean you want me to walk out of the courtroom?" Macaulay asked, with dramatized incredulity.

"Well, that's our feeling on the matter. Is it really necessary for us to be represented while people argue about concrete for weeks? I'm just disgusted with the district's position and I don't see why we should sit there supporting their case at our expense."

A weak speech, with that apologetic "just" and whiny

"I don't see why." You wouldn't catch Pericles saying that, I thought. Macaulay began to talk in the soothing manner that one saved for raving maniacs.

"Well, Dick, I can really sympathize with the way you fellows feel about this, but just walking out of the courtroom, I don't know. Let's see, if you could represent yourself instead . . . but god damn it, you're a corporation; you can't do that."

Macaulay was simulating serious exploration of my idea, even though I knew what the lawyers think about self-representation. ("Guy represented himself in San Jose on a purse-snatching charge," said Tony DeLuca over lunch one day; "first thing he asked the witness was, 'Did you actually see my face when I took your purse?'")

At this point, Macaulay was presumably facing one of the oldest scenarios in law: the client who wants to quit the game, leave it all up to the judge.

I can imagine that senior law partners have standard speeches they make to their young associates on that subject: "Now we know that clients sometimes become weary and disillusioned with the due process of law, and will occasionally even want to drop out of a trial," the speech would begin. "The trial lawyer must realize that such expressions reflect a state of emotional distress that is entirely natural to the client and should treat such outbursts with sympathy but firmness, presenting the client with the serious and even catastrophic consequences of proceeding without representation."

So Macaulay gave some respectful consideration to the suggestion, the suggestion that runs against the grain of the entire system. Then he said, "Dick, as soon as you leave the courtroom, you know what's going to happen. They're all going to turn on us."

There's the rub. It doesn't seem likely that the judge is so stupid that he is going to accept anything that's said about an absent party. But if a witness should say that I was drunk on the job, is the judge going to say to himself, "That sounds like a lot of baloney," or is he going to think, "If those guys refuse to stay in the ball game, then I'm going to accept whatever is said against them. That's the way the system works around here, right or wrong, and I didn't get to sit up here by bucking the system."

"Well, okay Duncan, why don't you go over there tomorrow and find out what's going on," I said, adding that we were prepared to go after the district's jugular vein, bring out how, back in the beginning, we appealed to the district to settle the issue with the contractor by negotiation, not litigation, how they told us that they had no quarrel with us and they didn't care whether the contractor sued them or not.

"Jesus, Dick, you didn't tell me about all this. Why have you been holding this back on me?" Macaulay was excited about that news, or pretended to be. He went on about how we were going to bring all that out when we presented our case. Suddenly I was struck by the bleak thought that the weeks of trial so far had represented only the contractor's case. The rest of us hadn't even begun. The trial would go on forever.

The next evening I was having dinner at a friend's house. Macaulay called, excited. The judge was pushing settlement, was going to ram $45,000 down the contractor Mullins's throat. Surely we all could come up with that, the judge had said, hinting that no one was clean in this situation.

"I told him you'd go $7,500, no more," Macaulay said. "The judge knows that you don't have any insurance. At least we managed to get that through to him. He asked me whether you could write a note for the rest of it. I said I just didn't think you'd be willing to go fifteen. But if we could go, say, thirteen, I think that would do it."

"And fifteen?" I asked.

"Fifteen and you'd look like heroes to the judge; it would look like you were really busting ass, working Saturdays and everything else to pay this thing off."

I wondered whether the other parties had exhausted their bargaining positions. DeLuca said he had, but was he holding more cards? Maybe not. Schultz said his insurance company's limit was $15,000.

"Consider yourself authorized to bargain up to fifteen," I told Macaulay.

The case was settled the next day, with my agreement to put up $13,000 cash into the settlement pot. Add to this $13,000 in legal fees, and $5,000 for my time, and my total cost came to $31,000.

Schultz, the structural engineer on the job, and his insurance company paid $15,000 settlement, plus, I suppose, $15,000 legal fees. Schultz himself had to put up only $5,000 cash, plus his time, but the insurance company will get back their money from him in the long run.

The district reportedly paid $35,000 in legal fees, plus $15,000 settlement costs, plus staff time—say $55,000 altogether.

Total litigation costs to the district and its consultants and their insurers, plus the contractor, his bonding company, and the landscape architects, were about $150,000.

The six law firms involved collected $88,000.

Mullins's net gain was $33,000, less the cost of lost business during the months of litigation and trial. I was told that he owed most of the $33,000 in back taxes. The lost business may not matter, for I understand that his contracting firm has been dissolved, and Mullins himself has taken a job as some kind of government inspector.

My father, like me, is a civil engineer. After the trial was over I told him the whole sad story, how I had collected $3,500 in fees on the job and paid out $31,000. He just shrugged, as if to say, "That's the world you people have created." My father spent his life building things. He's seventy-two now and still at it, fifty hours a week. He doesn't brood about professional liability or OSHA or affirmative action. "I'm an old man. What can they do to me?" he says.

I used to be a designer, and once I knew the secret and the satisfaction of transforming a concept into a reality. But that's dangerous business these days, and there is not much of a market for it, either.

But if you watch the lawyers, in time you learn the trick of trafficking in words. My firm has tripled in size since we worked on the San Ignacio project. We write environmental impact reports now, some of them so big it takes more than one strong man to lift them, and we earn a good profit performing studies and analyses required (but I suspect not read) by bureaucrats. "Forensic engineering" I call it. It's the new way, and business is booming.